DYNAMITE STORIES AND SOME INTERESTING FACTS ABOUT EXPLOSIVES

HUDSON MAXIM

Édition et mise en page : Culturea (Hérault, 34)
Retrouvez notre catalogue sur http://culturea.fr/
Contact : infos@culturea.fr
Imprimé en Allemagne par Books on Demand Gmbh
Design typographique et layout : Derek Murphy et Reedsy
ISBN :9791041985814
Date de parution : février 2024
Ce livre :
— a été composé avec une police open source libre de droits dessinée sur Open Fondry https://open-foundry.com/
— est édité en libre access sous licence CC0 (Creative Commons Zero)
afin de conserver l'oeuvre au plus près des caractéristiques du domaine public.
— offre la possibilité à son lecteur de partager, dupliquer ou distribuer son contenu, sans restriction ni réserve.
— permet de replanter un arbre. SHS Éditions soutient Plantons Pour l'Avenir, fonds de dotation pour le reboisement des forêts françaises
400 projets de reboisement de forêts soutenus depuis 2014 à travers la France https://www.plantonspourlavenir.fr/

INTRODUCTION

SOME INTERESTING FACTS ABOUT EXPLOSIVES

An explosive material consists of a combustible and of an oxidizing agent for burning the combustible. Hence it contains within its own substance the necessary oxygen for its combustion, so that it will burn without atmospheric air and therefore in a confined space.

There are two main kinds of explosive materials—high explosives and gunpowder. There are also two mains kinds of high explosives—dynamites and military high explosives. Lastly there are two mains kinds of gunpowders—black, smoky gunpowder and smokeless gunpowder.

Dynamite is used mostly for commercial blasting purposes, such as blasting rock in the construction of railways, and so forth. Military high explosives are mostly employed for submarine mines, warheads for torpedoes, and as bursting charges for high explosive projectiles.

A high explosive is consumed almost instantly by what is called a detonative wave; hence it is said to detonate. When gunpowder explodes, it is not consumed by a detonative wave, but burns from the surface, and the more strongly it is confined, that is to say, the higher the pressure under which it is burned, the more rapid is its combustion. Although the action is rapid, it is yet much slower than is the action of detonation of high explosives.

The name gunpowder is a misnomer, for gunpowder is no longer a powder, but is made in the form of hard and dense grains or sticks, according to the use for which it is intended.

A gunpowder is smoky when its products of combustion are not all gaseous. Only about forty-four per cent. of the products of combustion of black gunpowder is gaseous. The rest is inert solid matter, which makes the smoke.

The products of combustion of smokeless powder, however, are practically all gaseous. Consequently, weight for weight, it is much more powerful than black powder.

Black gunpowder is a mechanical mixture of charcoal, sulphur and saltpeter, the charcoal and sulphur being the combustible elements, and the saltpeter the oxidizing element or the element that supplies the oxygen.

In smokeless powder the oxygen is held in chemical union with nitrogen and hydrogen, but the bond between the nitrogen and the other elements is weak, so that when ignited the other more active elements are enabled easily to unite at the expense of the nitrogen.

In the combustion of all explosive materials, great heat is generated, and the force of the explosion is dependent upon the volume of gases and the high temperature to which they are raised.

The smokeless powder used in the United States is made by dissolving a special kind of guncotton or nitrocellulose in ether and alcohol, just sufficient of the solvent being used to gelatinate the nitrocellulose, which is then stuffed through a forming die into rods. The rods are cut into sections of about three diameters long. The die, the invention of the writer, contains seven mandrels arranged in such wise that when the material is forced through the die the bar is multi-perforated with seven holes at equal distances apart. The grains or rods of smokeless powder are then dried for use.

When burned in a cannon, all of the surfaces of the material are practically instantly ignited by a small flash charge of black rifle powder used for the purpose of setting fire to the charge of smokeless powder. The combustion in the perforations causes them to become larger and larger until the grain is all consumed. This form of grain tends better to maintain the pressure behind the projectile in its flight through the gun, and enables the use of larger charges of powder with lower pressures than could otherwise be employed. In fact, it would be impossible to use a smokeless powder made of pure nitrocellulose in big guns without the multi-perforations.

In certain European countries where the multi-perforated powder has not been adopted, nitroglycerin is employed, combined with the nitrocellulose, which causes the material to burn through a greater thickness in a given time. Thus a smokeless powder may be made without the multi-perforations, but smokeless powders containing nitroglycerin erode the guns and destroy them very quickly, while guns employing pure nitrocellulose smokeless powders last much longer.

When one of our big army or navy cannon is fired, the time which elapses from the instant of complete ignition of the powder charge to the instant that the projectile leaves the muzzle of the gun is about the fiftieth or the sixtieth of a second, and in that time the hard and horn-like smokeless powder material is burned through only about a sixteenth of an inch; hence the rate of combustion or rate of explosion of smokeless powder in a cannon is about four inches per second, while it has been ascertained by actual experiments that the rate of combustion or rate of explosion of dynamite and other high explosives is about four miles per second, so that the rate of consumption of smokeless powder, as compared to that of a high explosive, is as are four inches to four miles.

As the time required for the projectile to be thrown from a twelve-inch cannon is only about the sixtieth of a second, sixty of these huge guns could be placed side by side and fired by electricity one after the other, while grandfather's clock is making but one tick.

Our ideas of duration are but relative. We have seen that the combustion in a cannon, though very rapid to our senses, is actually very slow indeed as compared with the much more rapid combustion of a high explosive; and great as is the speed of the detonative wave, yet the speed of the earth in its orbit is four times as great.

If a celestial giant with a huge dynamite bomb the size of the earth itself were to approach the earth in its flight through space, and detonate the bomb immediately behind the earth, it would take half an hour for the bomb to explode, that is to say, it would take half an hour, or thirty minutes, for the explosive wave to pass through the eight thousand miles of its diameter. As the speed of the earth in its orbit is four times as great as that of the explosive wave, the earth would rush away, leaving the bomb about thirty thousand miles behind by the time it had completely exploded. If the interstellar ether were a high explosive mixture and were to be set off by the bomb, the earth would pass on clear around the sun, and while coming back, about six months later, would meet the explosive wave still going. It would require nearly a year for such a detonative wave to reach our sun from the earth.

We have seen that if the earth were a ball of dynamite, it would require half an hour to explode. If the sun were a mass of dynamite it would require about two and a half days to explode.

We frequently hear the theory advanced that planets and suns sometimes explode from pent-up forces within them, and that our earth might possibly blow up. Now, the force exerted by a high explosive is dependent entirely upon the pressure capable of being exerted by the gases liberated by the explosion. The pressure exerted by the most powerful high explosives has been estimated to be about 500,000 pounds to the square inch. Consequently, were the whole molten interior of the earth to be replaced with dynamite and detonated, the explosion that would follow would not lift the earth's crust. The superincumbent weight of the earth's crust is greater than would be the pressure exerted by the dynamite.

If it were possible to throw a projectile from the earth to the nearest fixed star, Alpha Centauri, it would take about four years for the light of the flash to reach that star. The sound, if it could travel through ether, would reach there about four million years later. The projectile, traveling more than twice as fast as sound, would reach there in about two million years.

When one of our big twelve-inch cannon is fired, the projectile, weighing a thousand pounds, has a muzzle energy, stated in mechanical terms, of about 50,000 foot tons, that is to say, its energy is equal to 50,000 tons falling from a height of one foot—energy enough to lift two 25,000-ton battleships to the height of a foot.

As the projectile weighs half a ton, the energy is equal to that which would be developed by dropping the projectile from a height of more than twenty miles, making no account of the resistance of the atmosphere.

Dropping upon a piece of armorplate too hard and thick for the projectile to penetrate, the heat developed would be sufficient to melt 750 pounds of cast iron.

When one of these projectiles is fired from the gun directly against twelve-inch armorplate, which the projectile is capable of penetrating, the hard-tempered steel plate in front of the projectile is fuzed or rendered plastic from the heat generated by the energy of the impact, and is forced like wax from the path of the projectile.

There are many popular errors regarding the action of explosive materials. One of the most notable is the opinion that the action of dynamite is downward, and that if a body of high explosive be detonated on the surface of the earth the main effect is downward.

The exact opposite is the truth. When a mass of explosive is detonated, it is converted practically instantly into a ball of incandescent gases and vapors under very high pressure. When confined the gases act to disrupt their container.

When a large steel projectile is charged with a high explosive, like picric acid, and the explosive detonated, the walls of the projectile are not only broken but they are also torn, twisted and shredded, and so quick is the action that the inner surface of the metal is compressed and densified against the outer metal.

For this reason it is easy to tell from the character of the fragments of a projectile whether or not a high explosive or an explosive of inferior power was employed, that is to say, whether or not the explosion was of high order or of low order.

There is one false belief about the action of high explosives that has been about the hardest of any to kill, and the cost of killing it has been very expensive. Furthermore, it possesses more lives than the proverbial nine-lived cat. This belief is that five hundred pounds or so of dynamite exploded upon a warship or upon coast fortifications would destroy ship or fortifications, and that a few of such large bombs of dynamite dropped in a city would lay the city in, ruins.

Upon the advent of the aeroplane and the dirigible balloon, it was confidently believed that the aerial bomb would quickly become the most destructive implement of warfare. It was prophesied that should war come between England and Germany, London would soon be reduced to a heap of ruins by bombs dropped from the German Zeppelins.

Several years before the European War broke out, I predicted that Zeppelin bombs would not and could not by any possibility work very wide destruction, and events have since vindicated my prediction. I pointed out the fact that should a hundred Zeppelins visit the city of London, once a day, for a year, returning to their base without mishap, and each Zeppelin succeed in destroying two buildings, the destruction would just about keep up with the growth of that city, for they build in London sixty thousand houses a year.

We all remember the destructive powers that were predicted for the fifteen-inch Zalinski pneumatic dynamite guns that were mounted at Sandy Hook and at San Francisco at enormous Government expense. These guns were capable of throwing with compressed air about six hundred pounds of nitrogelatin to a distance of from a mile-and-a-half to two miles. It was popularly believed that one of these bombs striking upon a huge armorclad warship would utterly destroy it.

Also two of these guns were mounted in a sort of cruiser called the Vesuvius. During the Spanish War the Vesuvius was taken down to Cuba, and in one action several of the huge bombs were thrown upon the earthworks and fortifications of the Spanish. They succeeded merely in mussing up the green, grassy effect. They did no material damage, for the reason that the action of the explosive was nearly all upward into the air.

When the pneumatic dynamite gun was promulgated, it was popularly believed that all high explosives were exceedingly sensitive, and that it was necessary to get them out of the gun very gently if they were to be thrown from ordnance.

The writer was the first to dispel this folly, through the invention of Maximite, a high explosive which will stand not only the shock of being fired from heavy guns at high velocities, but which will also, without exploding, stand the far greater shock of penetrating the heaviest armorplate—armorplate as heavy as the projectile will stand to pass through without breaking up.

While I was working upon Maximite and trying to get the Government to adopt it, Congress appropriated the money for building an eighteen-inch gun for testing a shell invented by Louis Gathmann, which was intended to destroy battleships by exploding the shell on the outside of their heavy armorplate, it being believed that if five hundred pounds of guncotton were to be fired against the side of an armored ship and exploded, the whole side of the ship would be blown in and the vessel destroyed.

The gun employed by Gathmann was essentially the same type of gun as that previously designed by me, and explained in a lecture by me before the Royal United Service Institution of Great Britain in 1897, and illustrated in a book of mine published the same year by Eyre & Spottiswoode, British Government printers, except that the bore of my gun, which was of the same weight as that of the Gathmann gun, was greater. With my gun, however, I proposed to throw armor-piercing projectiles, or projectiles capable of penetrating an object struck and exploding inside of it. I did not believe that a quantity of high explosive that could be thrown in a shell and exploded on the outside of a heavily armored ship would destroy it, but believed it necessary that the explosive should penetrate and explode inside the ship, and within earthworks and fortifications in order to destroy them.

Maximite was adopted by the United States Army in 1901. It was during that same year that the experiments were conducted with the Gathmann shell at Sandy Hook. I attended those experiments.

Two Kruppized armorplates, each eleven-and-a-half inches thick, sixteen feet long, and seven-and-a-half feet wide, and each weighing 47,000 pounds, were set up, one as a target for the Gathmann shell and the other as a target for the regular United States twelve-inch Army Rifle. Each of the plates was backed by supports to represent the same strength as though mounted on a battleship.

The Gathmann shell weighed about eighteen hundred pounds, and carried about five hundred pounds of guncotton, while the Government twelve-inch shell weighed a thousand pounds and carried only twenty-three pounds of Maximite. The Gathmann shell had a soft nose, which collapsed on the plate at the instant before the explosion of the shell, so that the guncotton might explode fairly against the side of the plate.

At the first shot of the Gathmann gun, the projectile struck the plate squarely and exploded, but the only effect upon the plate was to leave a great yellow smudge on its face. The plate was neither cracked nor pushed back. Several more shots of the Gathmann gun were fired, and although, under the heavy pummeling, the plate was pushed back and broken through, up and down, it was not otherwise injured.

Then the Government twelve-inch gun was fired at the other plate. The first shell contained nineteen pounds of high explosive, and it passed through the plate, leaving a clean round hole, and exploded behind the plate without breaking it. The next shell contained twenty-three pounds of Maximite, and the fuze was timed to go off a little quicker. This shell exploded in the plate when about two-thirds through, with the result that a hole was blown in the plate as big as a barrel, and the plate shattered into fragments.

One would think that these tests would suffice forever to seal the doom of the Gathmann type of shell. Nevertheless, it matters not what Army and Navy officers may learn by experience, or know without experience, Congress does not know and does not understand, and depends far more upon think-so than upon experience. The result is that Government officers are often compelled, as in the case of the Zalinski dynamite gun and the

Gathmann shell, to waste large sums of money while they know very well beforehand exactly what the results will be, and that the tests will prove the devices to be abject failures. Even after the failure of the Gathmann shell, another shell of almost identical conception and purpose was made and tested under a Congressional appropriation, to be relegated to the scrap-heap of failures.

It is very fortunate that things happen to be as they are in the cosmos and that the action of a high explosive when exploding against a massive body is to rebound from that body on the line of least resistance. It is for this reason that more damage is not done by great explosions.

One of the biggest explosions in the history of gunpowder manufacture occurred at Pleasant Prairie, Wisconsin, on the 9th of March, 1911, when it was estimated that a thousand tons of black blasting powder blew up. Glass was broken over a very wide area. Some glass was broken in Chicago, about fifty miles distant.

But neither the walls nor the foundations of buildings were greatly disturbed even but a few miles from the explosion. In the village of Pleasant Prairie, at a distance of but two miles, although the buildings were very much damaged the inhabitants continued to occupy them.

Early in the morning of July 30, 1916, a very large quantity, certainly several hundred tons, of high explosive materials blew up in New York Harbor, not far from Ellis Island. A large quantity of shrapnel ammunition and other ammunition went up in the blast, their fragments raining all over the surrounding water. There was but very little loss of life, and the actual material damage to buildings in Jersey City, Manhattan and Brooklyn was astonishingly small, except the loss from broken glass.

Why is it, then, that so much glass is broken and at such long distances, while the foundations and walls of buildings suffer but little injury? Let me explain. When a quantity of high explosive detonates, a wave of atmospheric compression is sent outward in all directions by the explosion. It is, in fact, a huge sound wave, and moves exactly at the speed of sound—about eleven hundred feet per second. Of course, buildings or other structures or objects near enough to the explosion to be struck by the expanding gases themselves, or by the atmosphere immediately propelled forward by them like a projectile, may be destroyed, but the area over which this action occurs is so circumscribed that no great damage is apt to result at distances beyond a few hundred feet.

However, the great sound wave may travel to a distance of many miles. Consequently, as a result of the explosion just referred to, about a million dollars' worth of glass was broken in New York City alone. One would naturally suppose that the fragments of window glass broken in this manner would fall inside a building, but they do not. Almost always they fall outside into the street. The reason for this is that the wave of compression, striking a pane of glass, forces it inward nigh to the breaking point, and then as the wave of compression moves on, followed by a partial vacuum, the glass, springing outward to fill the void, breaks, and falls into the street.

An interesting incident of this great explosion was staged at Ellis Island. There were a goodly number of immigrants on the Island at the time, congregated from the four corners of the earth, some of whom had come to America to seek their fortunes in this land of freedom-from-everything-except-freedom, but many had come to find quiet and security from war's alarums. Few of them, indeed, had ever felt the comfort of an overcoat, but many had dreamed of some happy day when they would sport a veritable fur-lined overcoat.

When the great explosion came it sounded like the crack of doom, and most of the immigrants believed it to be the real thing and proceeded with agitated precipitation to get their souls ready for rapid transit over the Great Divide.

All eyes naturally were averted to the celestial concave, aglare with the great conflagration, when suddenly, to the confounding amaze of all, a large flock of fur-lined overcoats began tumbling down out of the heavens all over the Island. It is true they were lined merely with sheep's fur, but even such a garment is as much the pride of the Northern European peasant as is the broad, glad-colored sombrero the pride of the Mexican peon.

As the Government statute books and rules and regulations governing immigrants contain no provision for the disposal of such species of manna as heaven-sent overcoats, the immigrants were the beneficiaries.

Great as are such explosions as that at Pleasant Prairie and that in New York Harbor, they are but little things indeed compared with the explosions that sometimes accompany volcanic eruptions. Mother Earth is the greatest of all explosive manufacturers.

Water seeping down into the earth's crust and trapped in large quantities in the neighborhood of volcanoes sometimes becomes heated to high incandescence—heated until it is no longer water or steam, but mingled oxygen and hydrogen, far above the temperature of their dissociation—under a pressure so great that they occupy a space no larger than the original water; consequently the entrapped waters exert a pressure as great as the strongest dynamite.

The most notable volcanic explosion that ever occurred in historic time was when that old extinct volcano, Krakatoa, in the Straits of Sunda, that had been sleeping for thousands of years, was literally blown into the sky by the pressure of the pent-up gases beneath it.

This great eruption occurred in 1883. More than sixty thousand persons were killed. The captain of a tramp steamer, who happened to be passing in the vicinity of Krakatoa at a distance of some miles, a short time before the explosion occurred, saw a very strange disturbance in the sea in the direction of the old mountain. Taking his glass he saw a perfect Niagara of water pouring into an enormous fissure that had opened in the earth. He was struck with consternation and rightly imagining that something very serious was likely soon to happen, he put on all steam to escape, and luckily he had reached a point which enabled him to survive the effects of the awful blast when it came.

The vast mass of water which had tumbled into the bowels of the earth was immediately trapped by the closing of the great fissure down which it had poured. The water was quickly converted by the intense heat into a veritable high explosive, with the result that the massive mountain was literally blown bodily skyward, and fell in huge fragments into the surrounding sea. The shock was so great that it was felt clear through the earth, and an immense tidal wave was set going which encircled the earth. The opposing portions of the great wave, meeting in the lower Atlantic, flowed up even to the coast of France. An atmospheric wave passed around the earth three times. It is estimated that the amount of volcanic mud that was discharged from the mountain during the eruption was more than the muddy Mississippi discharges into the Gulf of Mexico in two hundred years.

There was so much impalpably fine volcanic dust blown into the upper atmosphere that it did not entirely settle out of the air for more than two years, which period was noted for its beautiful glowing sunsets, due to the illumination of the fine dust suspended in the upper air.

As the ax is to the woodsman, so are high explosives to the engineer. With dynamite he hews down the hills, fills the valleys and tunnels the mountain-range to make a straight and even way for the locomotive. He cuts canals through the width of the land, uniting rivers and seas.

Always in the van of civilization, there is heard the churn of the rock-drill and the echoing crash and roar of the dynamite blast.

Also it is the huge high explosive shell that makes way for the march of modern armies, and high explosive mines and torpedoes are the terror of the underseas.

All forms of dynamite are high explosives, and all high explosives may fairly be called dynamite.

Smokeless gunpowder is actually but a modified form of high explosive. It is dynamite that has been chained and tamed by the chemist's cunning, so that it will burn without detonation, and thus permit the utilization of its awful energy to hurl shot and shell from war's great guns.

Thus it is that dynamite in its varied forms deserves the high place with steam and electricity as one of the great triumvirs that have been the architects of the modern world.

THE FORGOTTEN BIT OF FULMINATE

In experimenting with high explosives and in their manufacture, a little absent-mindedness, a very slight lack of exact caution, a seemingly insignificant inadvertence for a moment, may cost one a limb or his life. The incident that cost me my left hand is a case in point.

On the day preceding that accident, I had had a gold cap put on a tooth. In consequence, the tooth ached and kept me awake the greater part of the night. Next morning I rose early and went down to my factory at Maxim, New Jersey. In order to test the dryness of some fulminate compound I took a little piece of it, about the size of an English penny, broke off a small particle, placed it on a stand outside the laboratory and, lighting a match, touched it off.

Owing to my loss of sleep the night before, my mind was not so alert as usual, and I forgot to lay aside the remaining piece of fulminate compound, but, instead, held it in my left hand. A spark from the ignited piece entered my left hand between my fingers, igniting the piece there, with the result that my hand was blown off to the wrist, and the next thing I saw was the bare end of the wristbone. My face and clothes were bespattered with flesh and filled with slivers of bone.... The following day, my thumb was found on the top of a building a couple of hundred feet away, with a sinew attached to it, which had been pulled out from the elbow.

A tourniquet was immediately tightened around my wrist to prevent the flow of blood, and I and two of my assistants walked half a mile down to the railroad, where we tried to stop an upgoing train with a red flag. But it ran the flag down and went on, the engineer thinking, perhaps, from our wild gesticulations that we were highwaymen.

We then walked another half-mile to a farmhouse, where a horse and wagon were procured. Thence I was driven to Farmingdale, four and a half miles distant, where I had to wait two hours for the next train to New York.

The only physician in the town was an invalid, ill with tuberculosis. I called on him while waiting, and condoled with him, as he was much worse off than was I.

On arrival in New York, I was taken in a carriage to the elevated station at the Brooklyn Bridge. On reaching my station at Eighty-fourth Street, I walked four blocks, and then up four flights of stairs to my apartments on Eighty-second Street, where the surgeon was awaiting me. It was now evening, and the accident had occurred at half-past ten o'clock in the morning. That was a pretty hard day!

As I had no electric lights in the apartments, only gas, the surgeon declared that it would be dangerous to administer ether, and that he must, therefore, chloroform me. He added that there was no danger in using chloroform, if the patient had a strong heart. Thereupon I asked him to examine my heart, since, if there should be the least danger of my dying under the influence of the anesthetic, I wanted to make my will.

"Heart!" exclaimed the surgeon, with emphasis. "A man who has gone through what you have gone through today hasn't any heart!"

The next day I dictated letters to answer my correspondence as usual. The young woman stenographer, who took my dictation, remarked, with a sardonic smile:

"You, too, have now become a shorthand writer."

The grim jest appealed to my sense of humor.

On the third day I was genuinely ill and had no wish to do business. Within ten days, however, I was out again, attending to my affairs.

HELL SWAZEY BREAKS UP THE DANCE

About the first use of nitroglycerin in the United States as a blasting agent on a large scale was in the construction of the Hoosac Tunnel in Massachusetts, on the Boston and Albany Railroad.

So many accidents had occurred where the use of nitroglycerin had been attempted, that engineers and contractors were afraid to employ it. Nobel, however, had discovered that when nitroglycerin was absorbed in infusorial earth, it was rendered much less sensitive. This material he called dynamite.

A chemist by the name of Professor Mowbray believed that the main trouble with nitroglycerin had been that it was not sufficiently purified in its manufacture. He induced the builders of the Hoosac Tunnel to try his product. He built a laboratory on the side of Hoosac Mountain, over the village of North Adams, where he produced the stuff.

He put it up in tin cans, which held about a quart. For transportation these were carefully packed with cotton flannel between them.

The method of using the dynamite was to pour it into holes drilled in the rock, inserting an exploder cap and fuze in the usual manner. At that time it was popularly supposed that if nitroglycerin or dynamite were allowed to freeze, it became very highly sensitive and would explode on the slightest jar. Stories were prevalent that the sound of a fiddle string would explode nitroglycerin when frozen.

One day there came an urgent call from the east end of the Tunnel for more nitroglycerin. Professor Mowbray had in his employ a care-free and fear-free fellow by the name of Helton Swazey. When Swazey was sober, he was the soul of good nature, but when drunk, which was very frequently, he was as savage as a hungry cougar. This peculiarity earned Helton Swazey the nickname of Hell Swazey.

It was a very cold winter day when the call came, and Professor Mowbray, learning that Hell Swazey was going over the mountain that very evening to attend a dance, asked him if he would not take over the nitroglycerin with him. A hot-water bag was placed with the nitroglycerin and all was wrapped in a heavy blanket to protect it from Jack Frost. The shipment was placed in the back of Swazey's sleigh.

Hell Swazey's best girl, whom he took with him, did not know the nature of the cargo.

The nine-mile ride over the mountain was very cold. Swazey kept himself warm by imbibitions from a flask of liquid caloric, and to keep the young woman warm he took the blanket and the hot-water bag from the nitroglycerin for her comfort, leaving the explosive to the mercy of the below-zero weather.

When Swazey arrived at the dance-hall to join in the frolic, he was in so ugly and meddlesome a mood that he was promptly put out of the hall, followed by his woman companion. Swazey was mad all through. He went to the sleigh, and taking an armful of the cans of nitroglycerin, returned to the hall, and opening the door proceeded to hurl them with all his force at the merry-makers.

One can struck upon the stove and glanced across the room. Cans smashed against wall, ceiling and floor.

As the frightened occupants fled through the windows, they did as Mark Twain did when he saw the ghost—they did not stop to raise the windows, but they took the windows with them. In the language of Mark, they did not need the windows, but it was handier to take them than it was to leave them, and so they took them.

When Hell Swazey turned up for duty the next morning, Professor Mowbray had already heard of the escapade, but he was filled with marveling why the nitroglycerin had not exploded, particularly as it must have been frozen very hard.

When Swazey entered the presence of the Professor, he expected immediately to be discharged. He was meek and crestfallen enough, and began to excuse himself and to apologize for his behavior.

To his amazement, Professor Mowbray appeared to be very much interested and pleased, tapping his forehead with his finger, smiling and nodding, and muttering to himself, "Good; good; splendid!" He interrogated Swazey carefully, to be assured that the nitroglycerin was frozen hard, that it had been thrown hard, that it had struck hard, and that it had not exploded.

That very night there was mailed at the North Adams Post Office an application for a patent for freezing nitroglycerin to make it safe to handle.

THE POET'S UPLIFT

Explosive factories are veritable schools of efficiency. All work is done under the eye of the most vigilant caution, and the penalty for negligence is so expensive in the destruction of life and property that science, which is knowledge, and proceeds from sure premises to safe conclusions, is the sole guide. It does not do to follow a guess. The dynamite factory is no place for that class of persons who believe themselves to be favorites of Providence or of Almighty God, for dynamite plays no favorites.

There is probably no other class of persons so little guided by science as are the poets. They pride themselves on the fact that they ignore science. They claim that poetry is a sort of transcendental stuff, star-dusted from the gods' abode upon only a few persons fortunate enough to be born with a divine afflatus, which puts them into a fine frenzy—a condition of body and mind partaking somewhat of the ecstaticism of the Whirling Dervish, the spiritual clairvoyant and the soothsayer—a holy hysteria—a delirium-tremendous effervescence of over-soul—in which condition they are able actually to commandeer the co-operation of the Deity.

To heighten the humbug, the poets claim, to quote, that "poetry knows no law," that "it is above and beyond all law"; and consequently that it is "the antithesis of science," veritably "the despair of science," "defying all attempts at analysis and understanding," and that, being an inspired product, "poetry is the greatest achievement of the human mind."

The poets would have us believe that all of the great inventors and discoverers, scientists and philosophers, have been far inferior to the poets. The poets would have us believe that all the triumphs of chemistry and mechanics have been small compared with the triumphs of poetry. The poets would have us believe that the invention of the phonograph, of the telephone, of wireless telegraphy, the discovery of radium and the X-ray, the discovery of gravitation, are not equal to such triumphs of the poets as "Aurora Leigh," "Curfew Must Not Ring Tonight," and "The May Queen."

The poets would have us believe that the discovery of the spectroscope, which tells the composition of the stars so far away that the light by which we see them now left its source before the building of Babylon and the founding of the Egyptian Pyramids, is a less wonderful product of the human mind than is Shelley's "Skylark."

It is perfectly safe for the poets to live and move and have their being in error, but it does not do even for a poet, when working with explosive materials, to eliminate scientific procedure, for in that case he is likely to get an uplift that will sprinkle the feet of the angels with his filamented fragments.

This very thing actually once happened in the Pennsylvania oil region when the poet laureate of his community was blessed by the discovery of petroleum on his otherwise worthless farm. One well sunk by the oil company gushed a large quantity of both oil and natural gas. The royalty received by the poet was immense. One day he conceived the idea of climbing to the top of the oil-derrick and writing a poem to vent his pent-up fervor.

He had engaged the services of a photographer to catch his beatitudinations. The sun was just descending the horizon, and the poet and the top of the derrick were still aglow in the radiance of sunset, while derrick and poet were enveloped in an explosive mixture of gas and air a hundred feet in diameter. The photographer had said, "Beady, look pleasant, please." This was the moment of inspiration. The poet loosed his divine afflatus and set his fine frenzy to doing things. The following science-confounding doggerel is what he effused:—

Poetry is a divine art
And I am a poet to the heart,
And am writing these lovely lines
Right where the setting sun shines,
Just at the close of a beautiful day,
Under the milk-like Milky Way,
But which cannot be seen just yet though
Because of the sunset's brighter glow.
Yet I know it is there, and poesy may
Raise me nearer the Milky Way.

... And it did, for at this point the poet struck a match to light a cigarette, and the explosive mixture of natural gas and air about him fired first.

When last seen the poet was headed for the Milky Way.

HOW BENDER LOWERED THE PRICE OF DYNAMITE

Once, when entering my storage magazine at Maxim, New Jersey, in which were several carloads of dynamite, along with 37,000 pounds of nitrogelatin, made to fill an order from the Brazilian Government, I saw John Bender, one of my laboring men, calmly but emphatically opening a case of dynamite with cold chisel and hammer. With some epithetitious phraseology, I dismissed him.

It was not long after this incident, when the Boniface of the inn at Farmingdale, a nearby village, called upon me to buy some dynamite. He told me that he had employed John Bender to blow the stumps out of a meadow lot. I related to him my experience with that reckless person, and tried to impress him with the fact that Bender was temperamentally so constituted as to court death, not only for himself but for others about him, when handling dynamite.

But Boniface was unconvinced. He wanted Bender to do the work and he wanted the dynamite to do it with. Bender, he said, had assured him that he was a great expert in the handling of dynamite—that he could so place a charge under a stump that he could always tell beforehand the direction the stump would take, and about how far it would go under the impulse of the blast. Therefore, it was only a question of the price of the dynamite.

"Well," said I, "the dynamite you want is sixteen cents a pound, but I'll bet you the dynamite against the price of it that John Bender kills himself with it, so that if he does not succeed in blowing himself up and killing himself with the dynamite, you can have it for nothing. On the other hand, if he does blow himself up, you must pay for the dynamite."

A few days later, there was some hitch in Bender's exceptional luck. A particularly refractory old stump had resisted a couple of Bender's dynamic attacks. The failure to dislodge the stump Bender took as a personal affront, because it reflected upon his skill as a stump-blaster.

"Next time," said he, "something is going to happen."

He placed about twenty pounds of dynamite under the deep-rooted veteran, touched it off, and several things happened in very quick succession. The huge stump let go its hold on earth, and proceeded to hunt Bender. It was a level race, but the stump won. Striking Bender on the north quarter, it stove in four ribs, dislocated a few joints, and damaged him in several other respects and particulars.

Boniface came to settle for the dynamite.

"Sixteen cents a pound," I said. "Bender hasn't a chance in a hundred. Wait till the doctors are through with him."

"What do you say to a compromise," suggested Boniface, "of eight cents a pound? For really," quoth he, "I do not believe that Bender is more than half dead."

And the account was settled on that basis.

FOOLHARDY KRUGER

One of the most dare-devil men I ever had in my employ was a young fellow by the name of Joe Kruger. He was a very hard worker, and that won pardon for his many indiscretions.

I sent him one day to a neighboring explosives works to get a special kind of guncotton made there, and told him to have it sent by freight in a wet state. Instead, however, he filled about fifty pounds into a big burlap bag, in a perfectly dry state, and took it on the train with him and into the smoking-car, placing it on the seat beside him. He struck a match, lighted a cigar, and smoked throughout the entire journey. Had the least spark of match or cigar fallen upon the bag, the guncotton would have gone off with a tremendous flash and, although it would not have detonated, it would have burned him terribly, as well as any persons sitting near, and would have blown out all of the windows in the car.

At another time, in order to test the insensitiveness of a certain high explosive, a quantity of it was charged into a four-inch iron pipe, and the pipe hung against a tree as a target to ascertain whether or not the bullet would penetrate the high explosive without exploding it.

Kruger and I fired several shots with a Springfield rifle from cover at long range without hitting the cylinder of explosive. I was then called away and told Kruger to continue firing until he hit the mark. As soon as I left him, he advanced with the gun to within a few rods of the tree. His first shot penetrated the cylinder, exploding it with terrific violence, blowing the tree, which was about eight inches in diameter, clean off, while the fragments of metal flew about his head like hailstones. But none happened to hit him.

The following is the sort of adventure that is likely to happen to anyone under similar circumstances and has doubtless happened before and since.

Kruger had a dog which was well trained to fetch anything that his master threw for him. One day Kruger took some sticks of dynamite and went to a neighboring stream with the intention of dynamiting some fish. He attached fuze and exploder to a stick of the explosive, and threw it toward the stream, but, missing his aim, the dynamite landed on a rock.

The faithful dog, thinking that the stick had been thrown for him to bring, ran and returned with it to his master in great glee, with the fuze sizzing nearer and nearer to the explosive. Kruger ran in horror, the dog after him, deeming it great sport. The dog being the better runner, danced about his master. Finding it impossible to escape by running, Kruger climbed a tree with all the alacrity he could muster, and had just reached a vantage of safety when the dynamite exploded, and the dog—well, the dog was holding the stick in his mouth when it went off.

DISCHARGING PAT

A works foreman of mine who had been employed as assistant superintendent in another dynamite factory told me the following story:

He one day intercepted an Irish laborer who was taking a barrel, which had been used for settling nitroglycerin, down to the soda dry-house, with the intention of filling it with hot nitrate of soda from the drying-pans. The foreman scolded Pat roundly, and told him that, should he do such a reckless thing again, he would be instantly discharged. The foreman then went to the superintendent's office and reported the matter.

In the meantime, Patrick, utterly ignoring the injunction, simply waited for the foreman to disappear, then proceeded to the dry-house with the barrel and began to fill it with the hot nitrate of soda.

Over in the superintendent's office the foreman had just completed his narration of Pat's carelessness, when there was a thunderous report and a crash of glass, and Pat's booted foot landed on the office floor between them.

The superintendent dryly remarked, "Evidently, Pat is already discharged!"

LINES TO A LADY

Some years ago, when I was conducting experiments with detonators for my safety delay-action fuze, which was adopted by the United States Navy in 1908 as the service detonating fuze for high-explosive projectiles, I received instructions that a parcel of fulminate detonators, made at the torpedo station, had been received and were being held for me at Fort Lafayette, and I was told to go to the Brooklyn Navy Yard, whence I would be taken in a tug to the Fort for them.

After having procured the package, I concluded that it would be much more expeditious for me to take a trolley car home than to return by the tug. On entering the car and seating myself, I placed the package beside me on the seat, keeping my eye constantly upon it. It was, by the way, perfectly safe to carry if subject to merely ordinary handling, but it would not do to jump on it or to kick it about much, for, in that case, there might be some energetic results.

No sooner had I comfortably seated myself in the car than a huge, determined, militant-looking woman entered, brushing a few small men aside. Seeing all the seats occupied except the space where the package was, she turned and hurled herself backward and downward.

Her movements were so quick that I had barely time to throw my left arm firmly under her, and, although I am unusually strong, I had all I could do to support her enormous bulk. When she felt my arm beneath her, protecting the package, she was all the more indignant and determined to crush the package in order to teach me a lesson, and she glared upon me fiercely. I finally succeeded, by throwing my shoulder against her, in toppling her sufficiently to remove the package with my right hand, and then I let her down upon the seat.

I seldom wax poetical, and never permit myself to write verses to ladies when I am not sure that they will be gratefully received. But, in this case, I side-stepped a little from my usual course, and, taking my note-book from my pocket, wrote the following lines, which I folded up nicely, and when I arrived at my street, I handed the paper to Her Militancy:

Dear Madam, I'm an anarchist.
That package was a bomb.
I'm on my way
Someone to slay,
And this is really true—
I didn't want to waste that bomb
On just the likes of you.

HE SEPARATED

The freezing point of dynamite is about eight degrees F. higher than that of water. Once frozen, it remains congealed at temperatures considerably above the freezing point. When solidly frozen, it can be detonated only with much difficulty, and even then only with great loss of explosive force. Consequently, when conducting blasting operations in cold weather, it is necessary to thaw frozen dynamite before using it. The process is neither dangerous nor difficult if conducted with ordinary precautions, but it may be made full of peril by carelessness or ignorance.

A friend of mine named Roynor, when gold-hunting in Alaska, had as a partner a venerable prospector whose only known name was Andy. Andy was the dynamiter of the combination, as well as chief cook and dish-washer.

The old man used to utilize the oven of the cooking stove for thawing his dynamite. Occasionally, he would forget that the dynamite was there until it was heated to the danger point. These little inadvertencies at last strained the nerves of Roynor beyond the elastic limit. He remonstrated to his aged partner with all the epithetitious sesquipedalian terminology of which he was capable, but nothing in the way of language or dynamite had any terrors for the old man.

"Andy," said Roynor, finally, "if you are not more careful with that dynamite, we are going to separate, and we are going to separate the very next time you put any dynamite in the oven."

The following evening, as Roynor was returning from his day's work, and when nigh the shack where his partner was cooking, he saw the shack instantly convert itself into a blinding flash, which solidified into numerous scattered débris that flew by him and fell round him in abundance.

When he recovered from the stunning shock of the explosion and dazedly looked about him, he saw many fragmentary evidences of the repetition of the prospector's carelessness.

"Well, Andy," he sadly remarked, "I told you we should separate the next time you did it. We have separated all right—particularly you."

THE WELL-DIGGER'S CASUALTIES

At my laboratory near Lake Hopatcong, one of the natives, who had made a reputation as a well-digger, and claimed to be able to descend through more rock in a day than could any other living man, thought that his strenuous habitude would adapt him to the manufacture of explosive materials, and with this in view he applied to me for a position.

My foreman gave him a job in which his duty was to assist with the rolling of motorite. The foreman gave the fellow explicit instructions about the care necessary to keep his fingers from getting in between the rollers, as it would not only prove uncomfortable for him were he to shed a finger or a hand, but it would also spoil the motorite by mixing it with his lacerations.... Almost at once, the end of one finger went.

Immediately, the well-digger was discharged, for his own sake and for the sake of motorite.

The man next took a contract to dig a well for one of the cottagers on the Lake. It was in the early winter. The weather was cold, and his dynamite froze very hard. He placed it in a bucket of boiling hot water, which thawed the outer stratum of the frozen stick, overheating it and rendering it very sensitive, while the core remained frozen solid.

He was too active and impatient a workman to wait long for a stick of dynamite to thaw, so he took the partly thawed stick, seized a hatchet, and proceeded to chop off one end of it.

The blow of the ax upon the soft, overheated, highly sensitive portion, compressing it against the frozen interior, which served as an anvil, exploded the stick. There was one finger and the thumb left on his right hand which held the ax, while his left hand, which had held the dynamite, and his whole left arm, were blown away.

When he looked about him with the one astonished eye that was left, he seemed pained that his old friend dynamite had gone back on him in that way.

THE RIVAL EDITORS

The following story was related to me by a professional liar, and yet I have suspicions that it is not true in every detail; but I feel sure that some variant of it has been true more than once, with the exception of the aerial incident.

A certain inventor had invented one of the very often-invented high explosive compounds of chlorate of potash, sulphur, charcoal, paraffin wax, etc., thinking that he had made a great discovery.

Now it happens that there is so much erraticism about high explosive mixtures with chlorate of potash as a base that the pathway of invention of such compounds has been strewn with the wreckage of the hopes and anatomy of their inventors.

The inventor had enlisted the financial support of a promoter, and the promoter was endeavoring to enlist financial support for himself, and to that end had invited several men of means, with two rival newspaper editors of the place, to witness a demonstration of the explosive at the inventor's laboratory, which was a two-story, light frame structure.

The promoter was letting himself be interviewed by the two editors and other newspaper reporters on the upper floor, while the inventor was making a demonstration with some of the stuff on the lower floor, the prospective investors warily watching the proceedings from a respectful distance.

The inventor had about half a barrel of the stuff in a tub. He first took a portion of it and pounded it on an anvil to show that it would not explode from shock. Next he took a handful of it and threw it into the fire under the boiler, to show that it would not explode from mere ignition. He then took a hot iron, which he had brought to a white heat in a forge, and thrust it into the half barrel of the infernal mixture, to show that it simply could not be exploded except with a very powerful exploder or detonator.

But the mixture happened, on that occasion, to differ somewhat from the inventor with respect to the sequence of eventuations—and exploded.

The building went up, and the promoter, the two editors and the reporters on the upper floor accompanied the building.

Two of the newspaper men were great rivals. One of them was the editor of the Clarion and the other the editor of the Echo. It so happened that the Clarion had better facilities for getting telegraphic news than the Echo, and accordingly the Clarion was usually able to post its news in advance of the Echo, and the editor of the Clarion used often to chaff his rival with the remark, "It's no use to put up your poster now, for my poster of the same news is just coming down." He called the Echo the echo of the Clarion.

When the explosion occurred, the editor of the Clarion, being more directly over the explosive than was the editor of the Echo, went up farther and faster, and on his return met the editor of the Echo still going up, and called out to him, "Behind as usual! All of the other fellows are coming down."

THE PASSING OF "JEOPARDY"

We once had a servant girl whom we nicknamed "Jeopardy," because she could not be prevented from pouring kerosene directly from the can upon a lighted fire.

One day, Jeopardy left us very suddenly, and she never came back. We were sorry she left, as Jeopardy was a good girl. It developed that she had chanced to find a fifty-pound case of dynamite sticks in the wood-shed, which she had been using to start the fire in the kitchen stove.

Sometimes, dynamite will work all right for such a purpose, but it is notional stuff and can not be depended upon merely to burn. It was during one of these intervals of independability that Jeopardy went.

THE INVOLUNTARY ATTACK

Soon after the invention of the Maxim automatic machine gun, I took the American agency for the introduction of the weapon to the United States Government. Among the tests that were conducted with the gun at Sandy Hook was one known as the sand test, sand being sifted into the mechanism of the gun, which was then loaded and fired. The gun went through the test perfectly.

The commanding officer, however, had not himself been present at the regular tests and arrived upon the scene only after they had been concluded. This particular officer was a dyspeptic, and was at times very unpleasant and domineering. On this occasion, he was particularly so. When told by the officers immediately in charge of the tests that they had been concluded, he peremptorily commanded that the gun should be loaded and fired again. One of the under-officers demurred, stating that a sand test was a very hard one on the gun, and that it would be unfair to subject it to unnecessary hardship of that character. That officer was immediately sat upon very hard.

The gun was loaded and made ready, pointing out to sea, as usual. At this moment, a schooner was seen rapidly coming into range. The commanding officer, however, said that he wanted to see only a few rounds fired, and that there would be plenty of time to fire them before the schooner came into the zone of danger; and he immediately gave the command: "Fire."

My assistant, who was operating the gun, instantly obeyed. After the discharge of perhaps twenty-five rounds came the command: "Cease firing!"

But the gun kept right on. Then, the command came several times in loud shouts, but the gun did not hear. The rage of the commanding officer was at white heat, but it did no good. The gun kept right on firing.

There were three hundred and thirty-three rounds in the belt, the weapon had been rigidly clamped to a set direction, and my assistant, being a little bit rattled at the loud shouts of the commanding officer, did not think to unclamp it, and turn it out of range of the schooner.

Soon, a stream of bullets, flying at the rate of six hundred a minute, were ricocheting all about the schooner, and there was wild excitement and waving of hands on board—all to no purpose, until the last cartridge had been exploded.

The trigger had been pulled by the sand and held pulled. It was, consequently, impossible to stop the gun from firing, until the belt of cartridges was exhausted.

I felt glad. The subordinate officers also looked gratified.

HOIST WITH HIS OWN PETARD

Liquid nitroglycerin is still used to torpedo the oil-wells when they get old, in order to give them a new lease of life.

There was one teamster in the old days who had become notorious as a hauler of the dangerous explosive. The law does not permit the shipment of the liquid by freight or by express, and for that reason this teamster had plenty to do in hauling nitroglycerin for long distances. He was a great smoker and his old pipe was always alight, though he might be riding on a ton of nitroglycerin with a few kegs of black gunpowder chinked into the load.

One day he was carrying, on runners, about two tons of nitroglycerin and a few odd kegs of gunpowder, when something happened. There had been a fall of several inches of light snow the evening before, and the scene of the eventuation was an open field which he was crossing.

There was an enormous crater in the ground; the light snow around the crater was besprinkled with a few shreds of horse and harness and a sliver or two of sled, but not a trace of the driver was ever found.

THE FORGOTTEN PRECAUTION

I once hired board and apartments at the house of a Frenchwoman, who took in only a few select gentlemen boarders. Perhaps I may have been justly esteemed the star boarder, inasmuch as I paid the highest price, and, too, in addition to a sleeping room and a library, I hired another large room to serve me as a laboratory. Although my main laboratory was located at my factory, still I was in the habit of conducting a few experiments in a small way when not at the factory.

I had given my landlady particular instructions about not handling the various things in my laboratory. I strictly enjoined her not to touch anything under any circumstances—I would keep the place in order myself. Nevertheless, she could not be prevented from entering the laboratory to dust and tidy it up a bit, and she generally knocked over a thing or two in the process.

One day, I brought home a pint glass jar of pure nitroglycerin, setting it up out of reach of the little three-year-old girl, who often used the laboratory as a playground, in spite of my protestations. I called my landlady's attention to the fact that this bottle contained nitroglycerin, and I explained its dangerous character unless it were left undisturbed.

I told her that, if she found it out of the question to let the bottle alone, and should, in dusting, succeed in knocking it over and spilling its contents upon the table where it stood or upon the floor, and should wipe up the oily liquid with a rag, not to put the rag in the stove, for, if she did, she would blow the roof of the house off, and project herself into the empyrean, and through it and out at the other side.

She actually remembered this injunction for more than three days, but, on the fourth day, on my return home, the little three-year-old met me as I came in, and said:

"Mamma very sick. Cure Mamma."

"Mamma" was lying upon a sofa, pale as a ghost, and breathing heavily. When I asked her what the matter was, she answered, "Oh, I am so sick!"

I began to be thoroughly frightened, and wormed out of her the fact that she had a terrible nitroglycerin headache. It came out that she had been dusting and tidying the laboratory that day, and had inadvertently knocked over the bottle of nitroglycerin. Fortunately, it did not explode as it fell, the contents being merely spilled upon the table and floor.

She took an old towel and soaked the liquid up with it. She then rolled up the towel in a tight, snug, compact wad, and started toward the kitchen to put the wad in the cook-stove and burn it up, when, just as she arrived at the stove, she felt a dizziness in the head, and a strange sort of sinking sensation in the stomach. The top of her head began to buzz and pound.

Then, she saw light. It dawned upon her, like the inspiring flash that came upon Saul, that this was nitroglycerin, and she recalled what I had told her about the effect it would have upon her if she handled it, and my direction that, if she should spill the stuff and then wipe it up, she must not burn the rag.

THE FATAL HAT

Out in the Pennsylvania oil regions in the early days, while nitroglycerin in the liquid state was being used experimentally as a blasting agent, some boys found in a creek an old felt hat, which had been used as a filter for nitroglycerin.

One of the boys accidentally discovered that when laid upon a stone and the edge of the hat hit with a hammer, it would crack, so they took it to a blacksmith's shop, where they could have some fun by hammering it on an anvil.

At the first blow the old hat exploded. Two of the boys were killed outright, and two more were badly injured.

The blacksmith at the time of the accident, happened to be standing outdoors, which thereafter constituted his blacksmith shop until he could rebuild.

A DROP TOO MUCH

Professor Mowbray, who made the nitroglycerin for the Hoosac Tunnel and afterward served the American Xylonite Company many years as consulting chemist, conceived the idea that he could make a very powerful smokeless gunpowder by the use of nitroglycerin merely absorbed by fibrous guncotton and rolled into pellets. He had at the time a young assistant chemist at work for him, who has now become a man of much wealth and prominence in New York.

The assistant prepared some of the pellets under Mowbray's directions, loaded them into a rifle under wad and ball, and fired at a target made of several layers of pine boards. But the pellets did not seem to give the bullet the required penetration. Mowbray suggested remedying this defect by adding a little more nitroglycerin, which was done. The young chemist demurred a little. Still, he did as instructed—loaded and fired the piece again, with but little better results. This time, however, the breech mechanism stuck, and was opened with difficulty.

Mowbray said that there was but one thing to do, and that was to add a few more drops of nitroglycerin. It occurred to the young chemist that this sort of gunpowder came pretty near being dynamite, and he declined to fire the piece the next time, and was deaf to all entreaties of the Professor. As a compromise, the gun was rigged up on a rest, pointing at the target; a string was attached to the trigger, which the assistant, standing behind a barricade, pulled.

This time, there was considerable penetration of the target, and the walls of the building where the test took place were penetrated in many places, not with the bullet, but with the fragments of the exploded weapon.

Mowbray, hearing the report, ran out and ventured the suggestion that he guessed he must have got in a drop too much of nitroglycerin.

A CLOSE CALL

I had one very close call while conducting a sand test of the Maxim gun at Annapolis, where the Naval Proving Grounds were formerly located. The gun had passed through all of the regular tests satisfactorily, and it was then suggested to try if sand enough could be put into the mechanism box to block it and prevent its firing.

The gun fired perhaps fifty rounds before it stopped. Then it stuck, and my assistant worked at the belt and lever, attempting to start it again. I told him to put down the safe so that the gun could not fire, which he did. I was then about to step around the gun in front, which I confess was a very careless thing to do, when it began firing again. I was already so close to the muzzle that my clothes were cut by the bullets and burned by the gunpowder.

The trigger had been pulled, and held pulled, by the sand, so that the safe did not prevent it from firing.

It is pretty good practice to keep away from the business end of a loaded gun.

A PICKANINNY'S TREASURE TROVE

Once at Annapolis, while we were firing a six-pounder semi-automatic gun in a speed test, we had succeeded in firing forty-two aimed shots in a minute into a huge earth butt, which, owing to recent rains, was merely a heap of mud.

The day following, a negro boy, about fourteen years old, found one of the projectiles, which had penetrated the butt, and glancing, came out at the top without exploding. This he brought up to where my assistant was doing some work on the gun, and showed what he had found.

My assistant shouted at him, "Look out! That's loaded, and if you drop it, it might go off."

Frightened, the negro immediately dropped the projectile upon the hard cement pavement, and, as it struck point down, it did go off, and took off one of his legs; and a fragment of the shell came dangerously close to the head of my assistant.

NOT TO BE BUNCOED

The great Du Pont Powder Company had in its employ at one time a faithful, patient and lucky fellow, an Italian, who worked constantly, with not a day off except Sundays, for twenty-one years in the corning mill, breaking black gunpowder press cake into grains. During that period the coming mill had blown up seven times, once every three years, but each time Giovanni had happened, by the merest chance, to be outside for a few seconds to get a drink of water or on some other brief errand. Twice he had had his clothes nearly ripped off him, and his face and hands burned, such had been his proximity on these occasions to the crater of fire as the mill went up, and once he had been rendered unconscious by the shock.

Finally, at the end of twenty-one years of service, having put aside a snug little fortune, sufficient for the remainder of his life in sunny Italy, he packed up his belongings and turned his face toward his old home. Arriving in New York, his ticket purchased, he hied himself to a noted Italian hostelry, to await the coming of the joyous morrow when he should actually be on the big steamer, headed for home.

Giovanni had no bad habits, and the bunco man failed to lure him. He took no stock in the dapper, polished-mannered compatriot just recently from his home place, who was acquainted with all the folks. His cash was sewed into his clothes, and those clothes would not come off until he reached his destination.

When he was shown up into his room at night and left alone with his thoughts, a placard upon the wall above the gas-burner attracted his attention. It read: "Don't blow out the gas," and under this injunction was the statement that gas burned after ten o'clock would be charged extra.

Giovanni was indignant. Here he was at last caught between the horns of a dilemma. This, to his mind, was downright thievery. He would cut the Gordian knot. He would disobey the injunction. He would not pay for gas burned overtime perforce; and he blew it out....

An old sea-captain who had for forty years traveled on every sea, who had weathered a thousand gales, and survived a hundred shipwrecks, on his return from his final voyage, in making his landing on his home shore, slipped from the dock into the water and under the skiff, and was drowned.

Such is the irony of chance!

SIR FREDERICK'S BONFIRE

Sir Frederick Abel, who was the originator of the modern process of making high-grade guncotton and of compressing it into dense cakes for use, told me the following story:

At one time, Sir Frederick had about five tons of dry guncotton, which was not of sufficient purity to stand the Government tests. He had, on previous occasions, frequently demonstrated how compressed guncotton, though dry, would quietly burn away without exploding when ignited, so he now fancied that his five tons would make a capital bonfire. With this idea of entertainment in possession of him, he invited a party of friends to witness the unique conflagration.

The friends were dominated more by the spirit of aloofness than was Sir Frederick himself, and they kept at a respectful distance, while Sir Frederick advanced toward the pile of explosive, and threw a lighted torch upon it. Then he retreated a short distance to avoid the intense heat, for he expected to see the whole pile burn away.

It started by merely burning; but, as I have already said about dynamite, it is notional stuff. So, on this occasion, the guncotton took a notion to explode after it got fairly on fire, which did not take very long. The whole mass detonated with terrific violence, and, even before Sir Frederick had retreated as far as he expected to go, he was knocked senseless by the concussion, and nearly every shred of clothing was blown from his body.... Although considerably bruised and lacerated, he recovered after several months.

He had learned a useful lesson: that a small quantity of compressed dry guncotton can be very well depended upon to burn quietly away without detonating, but, when a large mass of it is ignited, the greater heat of combustion and the greater pressure generated in expelling the larger quantity of the products of combustion, is almost sure to produce detonation.

The fact that a small quantity of an explosive material will burn away quietly without exploding has often led persons to think that a large quantity would burn in the same manner.

At one time, the British Government had on hand at Woolwich Arsenal about a hundred tons of cordite that had begun to show signs of decomposition, and it was decided to burn it. The entire quantity was taken out into an open meadow, at what was supposed to be a very safe distance from the city limits. A train was laid to the pile and set on fire.

For the same reason that the five tons of Sir Frederick Abel's guncotton detonated, this huge heap of cordite also detonated. Almost instantly after it was ignited, it exploded with most awful violence, and with very disastrous results. A number of buildings in the near vicinity were leveled to the ground. A few persons were killed and many more injured.

THE IRREVERENT NATIVE

After I had sold out my interests at Maxim, the place was taken over by a dynamite-manufacturing company. As there was left in one of the magazines a considerable quantity of dynamite when the property changed hands, the new concern, not choosing to sell it as their own manufacture, proceeded to utilize it as fertilizer upon a field of potatoes.

One of the natives, with his team and helper, was engaged to do this work. They had been instructed to use great care in opening the cases, but they still held their own opinions about the care necessary, which were based largely upon the contempt that is born of familiarity, and, having arrived upon the potato-patch with a good, big load of dynamite, they began to knock the cases open in any old way.

There were no surviving witnesses, not even the horses.

AT FOLLY'S MERCY

After I had sold the works at Maxim and had invented motorite, I needed a place in which to make the material, and hired a branch of the works there for that purpose.

It was winter. My wife had accompanied me as a precautionary measure. She was sitting in the laboratory to keep warm, near a big barrel stove charged with bituminous coal.

On entering the laboratory for something, my wife asked me what was in those two tin pails sitting near the stove. She said that she had a suspicion it might be nitroglycerin, and she informed me that one of my men had just been in, stirring the fire, and that the sparks flew out in all directions, some of them lighting in the buckets, to be quenched in the very thin film of water floating on top of the oily liquid.

"Horrors!" I said. "It is nitroglycerin!"

I called the man who had placed it there, and told him to take it away. As it was necessary to keep the material from freezing, he took it into the boiler-house near by. A little later, on going into the boiler-house, I saw one of the men stirring the fire, while the other was standing with his coat-tails outstretched in either hand, forming a shield to keep the sparks from flying into the nitroglycerin.

It is practically impossible to make the ordinary man appreciate the necessity of care in the safe handling of explosives, and the life of the careful man is always endangered by the actions of the careless one.

THE WATCHMAN'S DOUBLE VISION

My successors in the use of the dynamite plant at Maxim had in their employ a day-watchman, an all-round combination useful and useless man, his usefulness and uselessness alternating with the alternation of his sobriety and inebriety.

One morning, after a night out, he proceeded to build the fire in the laboratory stove. To start up the kindling wood, he had been in the habit of lighting a handful of shavings, and then pouring on a little kerosene from a tomato can, which he kept upon a near-by shelf.

During that night, someone—possibly one of the laboratory operatives—had placed a similar can, filled with nitroglycerin, upon the same shelf, to keep it from freezing.

In periods of convalescence from his various stages of intoxication, the watchman had before seen two cans upon that shelf or shelves, but he knew that one of them was real, and the other an hallucination. Couldn't fool him that way!

Thinking that the hallucination would naturally be the lighter of the two cans, he took the one containing the nitroglycerin, and proceeded to pour it upon the fire.

There was so little of him left together after the explosion that, like Captain Castagnette, he died of surprise at seeing himself so dissipated.

THE ZEALOUS FOOL

On one occasion, at my laboratory near the shores of Lake Hopatcong, I was conducting some experiments to test the efficiency of the safety chamber of a detonating fuze for exploding projectiles charged with Maximite. The huge loaded shell armed with a fuze was placed in a pit and fixed so as to be set off by electricity from a distance.

To prevent any possibility of a circuit being formed to explode the detonator while making the connections at the pit, I went into the machine-shop, and opened the switch at the other end of the wires where they were connected with the battery. Not only did I take this precaution, but I disconnected also the wires themselves, in order to make assurance doubly sure.

Returning to the pit to connect up, my assistant, my wife and my father-in-law accompanied me. My assistant descended into the pit, while we stood over him, looking on. The instant he brought the wires in contact, the detonator went off. We looked at one another in amazement. It takes time to get thoroughly scared; but, as soon as we realized the full danger through which we had passed, we were numb with fright. Even now, when I think of it, I have a creepy feeling.

We had made half a dozen tests before this, and all of the shells had exploded except one. This was the second in which the safety-chamber had proved effectual. Had it failed this time, and had the Maximite charge exploded in the huge shell, we should all have been blown to ribbons.

I rushed back to the machine-shop, where I found that a certain employee—one of those careful, painstaking souls who are always attending voluntarily to the odds and ends of work left undone by others, had discovered the wires detached from the switch. With no memory of the rule that the switch should always be left open, he forthwith connected the wires, and then, to make his culpable industry complete, he closed the switch, thus making the electric connection with the loaded shell; and, doubtless, he was comforted by a sense of duty well done. His duties in my services certainly were done, for they ended right then and there.

SOME LIVELY COTTON WASTE

I once had an Italian laborer as man-of-all-work, who was rather a good-looking fellow. An exquisite mustache and a wealth of curly hair were sources of great pride and joy to him. One day he was engaged in burning up some rubbish, and to start a fire, took what he supposed to be a bunch of dry cotton waste, but which was in fact guncotton. Holding in one hand the wad of guncotton the size of his head, he applied a match to it. There was a quick, bright flash, and hair and mustache had disappeared. He did not mind the burn so much, but his anxiety about his appearance in the eyes of his sweetheart was pathetic.

SAVING TIME

When I had completed at my works, Maxim, New Jersey, a certain frame building of generous proportions, of which I was quite proud, and in which I had installed various processes and apparatus for making smokeless gunpowder, I told one of my assistants to have a gauge put on a large bell-drier that stood in a corner, which was employed for the time being to extract the moisture from about forty pounds of guncotton. He gave instructions to a machinist to do the job, telling him to remove the guncotton first.

As it was necessary for the machinist merely to bore a hole through the bell-drier and screw in the connecting pipe, he thought it a useless expenditure of time and effort to remove the guncotton. After he had bored the hole nearly through, he took a punch and hammer to knock out the remaining burr. A spark ignited the guncotton, and that bell-drier went right up through the roof and turned a somersault, striking about a hundred feet away. The walls of the building on the end where the explosion occurred were thrown outward, and the roof came down.

My assistant and another young man were in the building with the machinist at the time. Although dazed by the shock, they immediately rushed to the rescue of the poor fellow, who lay prostrate under a pile of burning débris. Not much could be done for the unfortunate, and he died soon afterward.

This instance is a type of many that result from inadequate precaution by workmen in the manufacture of explosives.

THE BROKEN SCALE

One of the closest calls that I ever had in my life occurred in my laboratory at Maxim, New Jersey, in the early nineties.

Two of my assistants and myself were weighing out small batches of fulminate of mercury from a ten-pound jar. There were on the bench as many as half-a-dozen small squares of glass, each with its little pile of fulminate upon it. There was also a five-pound bottle of nitroglycerin standing on the bench. A little way removed, and under the bench, was a fifty-pound can of gelatin dynamite.

We were proceeding very cautiously, when all at once the scoop toppled, and an iron weight fell, striking within an eighth of an inch of one of the pieces of glass on which was fulminate of mercury. After a second of suspense, we stared at one another in amazement, wondering whether or not we were still in the land of the living.

An investigation into the cause of the accident revealed the fact that one of the young men employed in the laboratory had broken off an arm of the scales—one of the supports of the scoop—the day before, and, with criminal reticence, had made absolutely no mention of the fact to anyone. Had that weight fallen upon the fulminate, it must have dealt death to all of us.

THE SINGULAR GOOD FORTUNE OF A GENTLE ENGLISHMAN

It so happened that during a tour of inspection seven of us were together, going over the works. On entering the guncotton dry-house, I noticed a strong odor of nitric acid.

"Out of here, quick!" I cried. "The place is going to blow up!"

There were perhaps a hundred pounds of dry guncotton in the room at the time, spread out in pans. As was afterward learned, the foreman, being in a hurry for the guncotton, had turned live steam into the pipes instead of circulating hot water through them as instructed.

We were barely out of the room when the guncotton burned with a flash, wrecking the building and setting fire to the fragments. I was just congratulating myself that no one had been injured by the explosion, when it was discovered that one of the party, an Englishman, the even tenor of whose way nothing could accelerate or disturb, and who feared nothing, had not quite made up his mind in time to get out of the room before the flash came. On seeing him emerge at last from the zone of destruction, I was horror-struck, for apparently every hair had been burned from his head and face, while shreds of skin hung from his hands and cheeks and brow.

Nevertheless, the Englishman's usual phlegmatic manner was wholly unruffled, and he spoke in his conventional voice, untinged with emotion:

"Mr. Maxim, it isn't often that one has an opportunity under such circumstances of witnessing exactly what occurs."

THE MATCH AT THE PEEP-HOLE

A certain patented device is used for the recovery of solvents in the manufacture of smokeless gunpowder. An acquaintance of mine conceived the idea that it would be an excellent thing to employ this same device for the recovery of alcohol used in the manufacture of felt hats. He conducted experiments successfully, having the hats placed in a chamber through which hot air was circulated, and from which it was afterwards conveyed to a refrigerating compartment to condense out the alcohol, then reheated and returned to the drying chamber.

Ultimately, this ingenious person so won the confidence of a company of hat manufacturers that they determined to build the apparatus at their factory, and to give it a thorough trial to test its practicability. Things progressed very well indeed, until there came a day when a leak was discovered in some part of the apparatus, and a plumber was called in to make the necessary repairs. This artisan's first act was to open a peep-hole, light a match, and peer into the drying chamber.

There was much instantaneity in the activities that followed. Fourteen persons were killed outright, including the plumber and his assistant, and the building was completely wrecked.

THE FLASK OF LIQUOR

Some years ago, in Austria, a worker in one of the mines found a flask nearly full of a liquor that he took to be whisky. Delighted with this treasure trove, he raised the flask to his lips, and gulped down a portion of the contents. Another workman, standing by, snatched the flask, and, in his turn, quaffed the liquor greedily.

That liquid in the flask was nitroglycerin, which, taken internally, is one of the most virulent of poisons. Both of these workmen were stone dead in less time than it has taken to tell this story of their fatal folly.

IMPERTINENCE PUNISHED

During the experiments at Sandy Hook which preceded the adoption of Maximite by the United States Government, a young lieutenant just out of West Point was placed in charge of the loading, although he knew absolutely nothing about explosives. He tried hard, however, to make up for his deficient knowledge by the most exacting, impertinent and foolish requirements.

I rebelled, but was told by the commanding officer that, while he fully appreciated the situation, he must, as a matter of duty, support his subordinate officer, and he advised me to return to my task in looking after the loading of the Maximite, under the direction of the impudent youngster. This I did.

The lieutenant, now having his own way, heated some Maximite very hot and filled a projectile with it through the false base plug provided for the purpose. There were two holes in the false base plug, through one of which the Maximite was poured into the projectile, while the other served as a vent. Being uncertain whether or not the projectile was filled solidly, the officer took a round stick, and rammed it down one of the holes, while he looked into the other. The result was that his eyes were filled and his face covered with the hot liquid Maximite, putting him out of commission for a week.

My sympathy for the fellow was quite overbalanced by my gratification.

CURIOSITY'S UPLIFT

Shortly after the Russo-Japanese war, there drifted in upon the Chinese shore one of the huge floating mines constructed by the Russians, containing about five hundred pounds of guncotton. This strange object greatly excited the curiosity of the Chinese, who flocked in large numbers to view it. While half a thousand of them were crowded in close upon the mine, marveling over the mystery of this flotsam, one of their number began to investigate it with a hammer, and, hitting the fuze a heavy blow, exploded the mine.

An American witnessed the event from a distance. Wondering what all the excitement was about, he had started toward the crowd with the intention of making an investigation on his own account, when, of a sudden, there was a flash and shock. The horde of Chinamen that had been clustered about the mine vanished in a cloud of dust. Fragments of heads, arms and legs rocketed skyward in the form of an inverted cone. The head of a Chinaman, severed from the trunk, went hurtling through the air, with the queue out-streaming behind, like a comet coming to perihelion. It passed just over the horrified American and struck the ground some distance beyond him.

PROUD EVEN UNTO DEATH

An inventor, who lived in the mosquito belt of Staten Island, constructed a dynamite gun out of a piece of four-inch-gas-pipe, and a dynamite bomb out of a short section of gas-pipe, capped at both ends. The bomb was filled with No. 1 dynamite. He placed several pads of felt between the projectile and the powder charge, to lessen the shock upon the bomb. By using small charges, he succeeded in firing a number of the projectiles safely. Although the velocity was low, still it was greater than that obtainable with the Zalinski pneumatic dynamite gun, which at that time was beginning to receive some measure of public attention.

The inventor was so fortunate as to have a "pull" with the congressman from his district, and through this influence he succeeded in getting Government permission for a test of his piece at Sandy Hook. In the meantime he had strengthened the powder chamber of his gun by driving on several steel hoops, in order to use larger charges of powder. So confident was he of the safety of his system of throwing high explosives, that, when the officers at Sandy Hook insisted on his retiring with them behind the bomb-proof during the firing of the piece, he balked and insisted that he be permitted to stand by his gun while firing it, as he had done in his previous experiments on Staten Island. He was not in the least impressed with any possibility of danger by reason of the fact that he was now using a much larger powder charge.

The officers, however, were obdurate. They told him bluntly that he must either stand behind the bomb-proof, or his gun would not be tested.

He replied:

"Very well, if Uncle Sam does not want my gun enough to let me test it in my own way, then I will sell it to foreign governments, and make Uncle Sam feel very sick and sorry."

On his return with his gun to Staten Island, he gathered together a party of neighbors and some representatives of the press, to witness the experiments that Uncle Sam had missed. When the gun was ready to fire, the little knot of spectators frayed out, and peeped from cover. There was but one shot, which was not a shot, but an explosion.

After waiting for some time for the inventor to come down and explain, the spectators went home, disappointed.

THE DOG THAT ATE DYNAMITE

In the early nineties I was experimenting with a new fulminate compound as a detonator for fuzes in high explosive projectiles. The compound consisted of fulminate of mercury with gelatinated guncotton and nitroglycerin.

One of my workmen had a pup of a miscellaneous breed, which would eat anything under the sun that he could masticate, and when anything was thrown into his mouth not too big for him to bolt, he swallowed it without the formality of chewing it.

One day his master gave him about half a pound of this fulminate compound. Another of the workmen put some metallic sodium and dry fulminate into a gelatin capsule, stuck this into the end of a quintuple dynamite cap, wrapped the whole thing in a piece of meat, and, calling the dog out into the field, made him stand up and "speak" for it. Then he dropped it into the dog's throat and it was swallowed at a gulp.

The next instant, the latter workman's own dog, which he prized very highly, came upon the scene and entered into a very brisk wrestling-bout with the dog that had been charged. Before he could call him away, there was a terrific explosion, and both dogs vanished from this vale of tears.

INSECURE SECURITY

Before the discovery by Nobel that the absorption of nitroglycerin by infusorial earth rendered it much less sensitive to shock, numerous attempts were made to bring it into general use, in liquid form, as a blasting agent, the most notable of which was during the digging of the Hoosac Tunnel. But, owing to its highly sensitive character, fatalities were numerous; while, furthermore, the necessity for perfect purity in order to render nitroglycerin stable—that is to say, to make it keep well—was not at first recognized, and many disasters were the result from explosions due to its decomposition.

The first attempt to introduce nitroglycerin as a blasting agent into the United States was made by a young German student. He called the stuff "glonoin oil." He brought over a few hundred pounds in cans on the steamer with him, most of which he disposed of during his sojourn in the States. But his venture was not a financial success, and he was obliged, when he returned to Europe, to leave an unpaid board bill at a New York hotel where he had been staying. He had left one fifty-pound can of glonoin oil, which he let the hotel proprietor hold as security, but which, however, later developments proved to be insecurity.

The glonoin oil occupied a place of honor in one corner of the barroom for several months after the departure of the German student. Decomposition having set in, yellow nitrous fumes began to emerge from the receptacle, the malevolent odor of which was soon noticed by one of the guests, who called the landlord's attention to the fact. Highly disgusted, the landlord picked up the can, walked to the front door, and threw it into the middle of the street. The act resulted in a miniature earthquake, which shattered the walls on both sides of the street, broke window-glass over a square mile, and landed the hotel proprietor in the hospital.

THE LOADED CHINAMAN

During the Russo-Japanese war a certain officer of the Czar, who was an impatient, overbearing person and a great martinet, had a Chinese servant whom he treated with the utmost harshness for the smallest delinquency, or for none at all. One of his favorite methods of inflicting punishment for offenses was to order the Chinaman to leave his presence, and, as the fellow went, to give him a hard kick.

The Chinaman aired his grievances one day to a Japanese spy, whom he took to be a brother Chinaman. The Jap suggested padding the seat of the Chinaman's trousers to prevent further contusions, and this was done, the padding being furnished by the Jap. A rubber hot-water bag was filled with absorbent cotton containing all the nitroglycerin it would hold. A small exploding device armed with percussion caps was placed in the bag so that the nitroglycerin would be exploded by any sudden blow. The unfortunate Chinaman was wholly unaware of the nature of the padding.

At the next meeting of the Russian with his servant, the poor Oriental inadvertently spilled some tea upon the officer's new uniform. Thereupon the enraged master proceeded to dismiss the Chinaman from his presence in the usual way, but with somewhat more precipitation.

One of the officer's legs was blown off, one arm was crushed to pulp, four ribs were broken, and it was more than a day before he was restored to consciousness. When he did come to, he found himself a prisoner in a Japanese hospital, having been left behind by the retreating Russians.

As to the Chinaman himself, poor fellow, he never knew that he had been loaded.

LIVING BOMBS

An American reporter, who was with the Japanese during the Manchurian campaign, told me the following story:

Column after column of Japanese had assaulted a Russian position, the capture of which was exceedingly desirable. Line after line of the brave little fellows was swept down by the unerring gun-fire of the Russians, but each time a few Japanese would scale the works, and go over them, only to be slain by the Russians inside.

There was a lull for a short space, and the reporter thought, as doubtless did the Russians, that the Japanese had given up the task, when, suddenly, a troop of perhaps a hundred Japanese rushed forward, in a widely scattered line. Onward they flew toward the Russian position, and, as they went up, there was a blaze of the Russian rifles, and half the Japanese column disappeared with a flash and a tremendous report.... They had exploded!

Each of them had been loaded with an infernal machine, hung across his breast and over his shoulders, so that should but a few of them reach the enemy's position, they could explode themselves and hurl death and destruction all around them.

The Russians were so astounded, so paralyzed by the spectacle and by the unexpectedness of it, that they ceased firing, while the remaining living bombs scaled the ramparts and leaped in among their enemies, who instantly vacated the place, flying like rats from a sinking ship.

SHIPS THAT PASSED IN THE NIGHT

During the Russo-Japanese conflict, more than one of the Czar's warships disappeared in a night cruise without leaving a trace.

I got the following story somewhat indirectly, and for that reason cannot vouch for its truth. It was told to my informant by a Japanese officer while in rather a more communicative mood than is usual for an officer of the Mikado.

This Japanese official at the time commanded a torpedo-boat. In the flotilla of which his vessel was one there was a torpedo-boat that carried neither guns nor torpedoes, for she had been stripped of all armament and every mechanical device not absolutely essential to her navigation, in order to lighten her. And then she was loaded with dynamite to her full capacity.

The Japanese officer declared that when a volunteer crew of half a dozen men was called for to navigate her, ten times the required number offered themselves, although they well knew that they were going to certain death.

The flotilla was steaming slowly through the darkness one night, not far from Port Arthur, when there suddenly loomed up ahead the huge bulk of a Russian warship. At once the dynamite-laden craft threw herself directly in front of the oncoming leviathan.

Without the pause of an instant, the doomed Japanese crew sprung the huge mine, when a vast cone of flame shot up, reddening the night, carrying with it high into the air, decks, superstructure and guns of the warship. The warship's magazines, fired simultaneously by the dynamite blast, aided the complete demolition. The returning torrent of guns and wreckage plunged into the sea.

All was over, and it was dark again.

A WILD PROJECTILE

In spite of every precaution at Government proving grounds, big projectiles do sometimes glance out of the butts or heaps of earth into which they are fired, or from the face of armorplates against which they are directed, and finally land in most unexpected quarters.

One day, while a thirteen-inch gun was being tested at Indian Head, a projectile glanced out of the butt, mounted high into the air, and then came down through the roof of a building, where there were engaged a number of officers and book-keepers. The projectile passed down through the floor, close to the desk of one of the officers, and buried itself in the earth.

As the projectile contained no explosive charge, the damage was not great, but the scare that was thrown into the occupants of that room was of considerable magnitude.

THE BOMB AND THE TRAIN

One of the most anxious moments that I ever experienced was during some experiments made by me at Maxim in throwing aerial torpedoes from a four-inch cannon.

These torpedoes were about four feet in length, charged with a very powerful high explosive, and armed with a detonating fuze. We had successfully fired several of them into a sand-butt where they exploded with great violence. There were six of them: five had been fired and the sixth was loaded into the gun, ready to be discharged, when a passenger train on the Jersey Central Railroad hove in sight, and was passing us about a thousand feet away as the gun was fired.

We had no idea of there being any danger to the train, as its position was well away from the line of fire, and each of the preceding projectiles had behaved so well. But, this time, the torpedo glanced from the sand-butt, and went after that train. We stood paralyzed with dread as we saw it pass over the train, close to the roof of a car, and strike in the swamp just beyond it, perhaps a couple of hundred feet behind the track. An inverted cone of black earth shot up, followed by a dull sound.

In imagination we had witnessed a frightful catastrophe, the wreck of a passenger train, with fearful loss of life, and all the horror of our own resultant predicament. Now that the danger was past, the even tenor of our way did take on a new relish. What objects we are, after all, of the mercy of chance!

THE MISSING VESSEL

At the same place, I was one day on the point of beginning an experiment, for which I required a small quantity of dry guncotton. Before going for it to the guncotton dry-house, I instituted a search for a suitable vessel in which to carry it. For some mysterious reason I had much difficulty in finding anything just adapted to my purpose, and the hunt delayed me a matter of five minutes or more. Finally, however, I secured a satisfactory vessel, and hurried out of the door in the direction of the dry-house.... I had covered less than a rod of the distance when the dry-house blew up.

In this instance, surely, a benign Providence interfered to save me from destruction.

THE DRUNKEN MESSENGER

Some years ago, soon after I had built my experimental laboratory near Lake Hopatcong, a dear old friend came to visit me. He had seen hard times in the interval that had separated us, and had suffered from both business reverses and ill health since the days when he and I were chums. He was plunged in the depths of pessimism, while I was optimistic. He was in the throes of abject discouragement. Though I made him many offers of assistance in varied forms, none of them seemed to cheer him in the least.

When I knew him in our youth, he had been one of the bravest men that ever lived; now, he appeared to have lost all his former courage. Often, however, he made the remark that he was minded to make an end of everything, since life offered him nothing worth while. I frequently importuned him against the folly of contemplating suicide.

It came about that one day I was in need of fulminate of mercury. As this material cannot be taken upon a train or sent by express, it was necessary to go for it with horse and wagon. Both my assistants and myself were just then too busy to be spared from the work in hand. So, it occurred to me that my old friend would be exactly the person to send on the quest. Since he was even then engaged in meditating suicide, he would not be in the least afraid of fetching the stuff for me. Of course, I should not have thought of sending him had I believed there was any particular danger. Certainly there would be none if the material were handled properly, and in a wet state.

My old friend started on the mission valorously enough, but he lost his courage presently, and returned empty-handed.

I then sent one of my helpers, a spare man who worked for me occasionally, as he had been long connected with the manufacture and handling of explosives. I gave him the necessary money for the purchase of the material, for the hire of the team and his other expenses, and as there would be two or three dollars over, I told him he could spend that in any way he liked, for his own use and behoof.

He returned along toward evening, left the horse and wagon at the stable, and started up to the works with a bag containing ten pounds of fulminate, placed in a small hand-valise. Fortunately I saw him coming soon after he had abandoned the vehicle.

The road was altogether too narrow for him; the ground seemed to reel under his feet, and he was steadying himself by swinging the valise back and forth from side to side with great violence. A drunker man never walked.

I took the valise away from him, and carried it to the works myself. The next day, when we opened it, we found that instructions had not been followed about wetting the fulminate. The bag of dry fulminate had, when he procured it, been merely set in a pail of water for a few minutes, and only long enough to wet a thin stratum of the explosive, leaving the whole interior perfectly dry.

It is surely a wonder that the drunken man had not exploded this mass of dry fulminate in the rough handling he had given it. Had he fallen with the bag, he must almost certainly have caused an explosion by the shock of the impact of the fulminate against the ground.

NITROGLYCERIN BY AUTOMOBILE

At another time, I required some very pure nitroglycerin. As this material, like the fulminate of mercury, could not be transported by either freight or express, it was necessary to go and bring it over by horse and wagon, or by automobile. I decided to go and fetch it myself with my automobile, which, at that time, was a Haynes-Apperson of one of the early makes. That machine had the faculty of going wrong oftener and in more places than any other piece of machinery I ever saw or heard of.

It had a very short wheel base, and, as the steering gear had worn so much that there was a good deal of lost motion, it required very great skill to keep the car in the road. No sooner would it be brought in line with the highway than it would immediately proceed toward the ditch or some wall or tree. It swayed from side to side of the road like a drunken man, and it was necessary, in order to keep it in the road at all, to calculate an average with it.

As it was quite a long drive to where I was to procure the nitroglycerin, I went into New York and brought out a young chauffeur from the Haynes-Apperson Company, explaining to him fully why I wanted him, and for what I was going. I made him understand that I did not want him to start to accompany me unless he had the courage to stand by me to the end. He was all courage; bravery seemed to ooze out of him at every pore.

When we started on our journey, early the next morning, I found that he was wholly unable to steer the automobile. He could not keep it in the road at all, and I had to drive it all the way myself; but, as he understood the machine and how to repair it, I concluded that he might prove valuable on that account. And he did, for during the outgoing and return trip, that old machine broke down three times, and the tires went flat four times.

On arrival at the factory, I let the chauffeur wait while I went to procure the nitroglycerin. I took a lot of bicarbonate of soda with me, with which I absorbed the nitroglycerin, forming a sort of paste. This rendered it safe to handle, and, by placing it in water, I could at any time dissolve out the bicarbonate of soda, and leave the pure nitroglycerin.

When I had prepared fifty pounds of nitroglycerin in this manner, placed it in glass jars and rolled them up with several thicknesses of felt covering material, I had them taken up to the automobile and placed in the rear part of it.

I then told the young chauffeur that I was ready to proceed, but he said that he had been talking with the men in the office, and that they had told him that they would not ride with Mr. Maxim in that automobile with that nitroglycerin for all the money in the world. They had frightened the fellow nearly out of his wits. It was with much persuasion and reasoning and insistence that I finally got him to consent to get into the car with me and ride along a very smooth even road to the skirts of the town, letting him believe that he could there escape, and that I would proceed alone.

When we got a little out of the town, I reminded him of his agreement to stick with me, and told him that it would be out of the question for me to attempt to proceed alone without an assistant, as I had but one hand, and could not repair the machine very well if anything should go wrong. But he was deaf to all entreaties.

Then I told him, with highly colored emphasis and significant gestures, that, should he not proceed with me, as he had agreed, I might prove then and there more dangerous to his comfort and well-being than the nitroglycerin—and I kept him with me!

After having traveled a few miles, the chauffeur began to recover his courage, and I had no more trouble with him.

As I was ascending a steep grade along a narrow road, on the return trip, I saw a big touring car bearing down upon me, with a party of four young men and two young women in it. They were traveling like the wind. I turned out of the road as far as I possibly could, and stopped my car, and signaled with my hand to them to slow down, pointing to the narrowness of the road.

They gave little heed to this, and rushed by me like a tornado, coming so close that they could not have missed my machine, hub to hub, more than an inch.

There is little consolation in the fact that, had they struck us, they never would have known how foolish a thing they had done.

THE JETS OF BLUE

A chemist friend of mine once invented a process of converting nitro-benzole into tri-nitro-benzole by a very quick and labor-saving method, which consisted in mixing the nitro-benzole with nitric acid, confining the mixture in a large, strong, steel cylinder, then gradually heating the cylinder until the required pressure should be produced, which was expected to effect the desired reaction.

Accordingly, five hundred pounds of nitro-benzole was mixed with the necessary quantity of nitric acid of the requisite strength, and the heating process was begun.

While anxiously watching this infernal machine, my friend saw a peculiar blue flame emerge from the seam around the head. Being of an alert nature, and able to take a hint without being kicked by an elephant, he withdrew from the vicinity of that cylinder. He did not merely sidle away from the perilous place—he fairly flew with an alacrity born of desperation. He had barely emerged from the laboratory when there was a terrific explosion that leveled the building, and formed an enormous crater in the earth where he had stood, the concussion knocking him senseless. And, today, he still swears, with solemn earnestness, that a freight car could have been buried in the hole that was blown in the ground when his pet project went off.

This experience so impressed him that he concluded that explosive compounds possess properties which place them in a class by themselves, and that it is a good class to avoid.

THE WISDOM OF RETREAT

During the experiments with Maximite at Sandy Hook, previous to its purchase and adoption by the United States Government, I was loading some shell in a small house, near where a ten-inch gun was being fired. About a year or so before, when a ten-inch gun of this particular make was being tested at Sandy Hook, the breech block blew out, went through the bomb-proof, and killed several officers and men.

Having completed my work, I started up the railroad track in the direction of the steamboat landing, to return home, when there came the ring of the bell for another discharge of the cannon. As the breech of the gun was pointing in my direction, I recalled the above fatality; but I reasoned with myself that, as a large number of tests had since been made with these guns, and as the defect in the breech mechanism was supposed to have been corrected, the chance of the breech blowing out at this particular discharge and coming my way was infinitely small.

Nevertheless, I thought, it is exactly on such occasions that the unexpected does happen. So I ran for all I was worth up the track and off at one side. Then I heard the report of the gun, and looking around, saw that sure enough the breech block had blown out, and saw it pass through a building in its path. I saw it strike the track over which I had been walking, cut a rail off, saw it strike the old stone fort beyond me, and ricochet high into the air.

There was an immediate shower of stones and débris falling around me, which I dexterously dodged. On examining the small house where I had been charging the Maximite shell I found that the windows were fairly riddled with pieces of smokeless powder, which had been blown from the breech of the gun.

THE RACE WITH DEATH

Among the many dynamite plants that hang upon the verdant hills of the American countryside, there is one which stands somewhat apart from the railroad, and the dynamite has to be carted to the station over the highway. At one point the highway passes close to the edge of a precipice of considerable height, at the bottom of whose abrupt, ragged sides nestles a pleasant villa, owned by a wealthy business man.

A friend of mine, who told me the story, had just paid a visit to this factory of explosives, and was walking leisurely along the road. At a distance of perhaps a hundred yards ahead of him there was one of the dynamite wagons, moving two tons of dynamite to the railroad. The driver had recently purchased a couple of fresh horses, which he pronounced "a spanking pair." They were rather restive and shied at everything they saw. But the driver was a brave fellow and a strong one, and he had no fear of being unable to control them.

All at once, under the impulse of a gust of wind, a newspaper flared up in front of them. Quick as a flash, they bolted, rushing headlong, the bits held firmly between their teeth; while the high-piled load of dynamite swayed from side to side menacingly as the wagon took the curves of the road.

At this instant the foreman of the dynamite-works flashed by, driving a pair of horses to an empty wagon. He had observed the plight of the driver of the dynamite wagon, and was lashing his horses in mad pursuit.

Although the foreman's team was inferior, still his wagon was empty, and he was soon neck and neck with the runaway horses. For several hundred yards it was a close race, neither achieving any appreciable advantage over the other. Nearer and nearer were they coming to the precipice, which yawned just where the road turned sharply to the right. Still on and on they flew, when, in a moment of advantage, the foreman leaped from his wagon, full upon the neck and head of the nigh horse of the runaway pair, and brought the team to a standstill within less than fifty feet of the precipice, and directly over the villa I have mentioned.

Had not this foreman possessed both the presence of mind and the athletic qualifications necessary, coupled with great daring, that load of dynamite must inevitably have gone over the precipice as the horses struck that curve. Little the peaceful occupants of the villa under the hill imagined what a calamity at that fearful moment overhung them!

THE INDOMITABLE POET

An editor in a large Western mining city once hit upon a happy expedient for getting rid of obnoxious callers. To this end, he filled a gunpowder keg with ashes, inserted a fuze, piled a handful of black gunpowder around it, to give the whole an air of reality, and established the arrangement on a table in his ante-room. On the advent of certain bores, the office boy followed instructions by lighting the fuze, and walking out of the room with the audible remark:

"I'm goin' to blow up that old guy in there!"

The thing proved its worth as an automatic bouncer, until, on a memorable day, a long-haired, calf-eyed, dreamy-looking young male person came into the place, who informed the office boy that he desired to see the editor. He explained in cadenced speech that he deigned to exhibit to the editor a poetic effusion, the lucubration of a fine frenzy, fairly oozing divine afflatus, on the Surplusage of Over-Soul in Young Maidens.

On hearing his minion's report concerning the visitor, the editor told the boy to light the fuze and to ask the poet to sit down; that the editor would see him in half an hour.

When the editor went out into the ante-room the fuze had burned out, the surface gunpowder had flashed off, but the poet was still sitting there.

SCATTERED

I was once called as an expert to visit a dynamite plant where a new kind of high explosive was being manufactured instead of the ordinary nitroglycerin dynamite. It consisted of a mixture of chlorate of potash, sulphur, charcoal and paraffin wax. Its inventor had given it the reassuring name of Double X Safety Dynamite.

A quarry-man in a nearby town had, with his safety-ignoring habitude, attempted to load a hole with the stuff, using a crowbar as a rammer, with the result that he set off the charge, and the crowbar went through his head.

This unscheduled eventuation aroused the apprehension of the president of the company, who was also its backer. He began to grow suspicious about the safety of the material. Being so much interested, he went with me on my visit of inspection.

We left the train at a siding about a mile from the works, and had just started in their direction when there came a sudden boom and roar, and the earth shook. Over the powder works there rose a huge column of black smoke, flaring wide into the sky.

We found a great crater where the mixing house had stood. Three men were working in the building when the explosion occurred. A fortunate survivor who had left the place a moment before to go for a bucket of drinking water, was walking about the crater, apparently searching for something among the scattered remnants. As we approached him, he sadly said:

"I can't find much of the boys. I guess you'll have to plow the ground if you want to bury them."

A LIVELY DEAD ONE

Several years ago, at the works of the American Forcite Company, a batch of nitrogelatin blew up in process of manufacture and several men were killed. One laborer who was working so close to where the explosion occurred that his clothing was nearly all blown off and he was spattered with the blood of his companions and crazed by the shock, started in a wild and aimless run along the road, with his tattered garments flying in the wind.

A woman of the neighborhood, whose husband was employed at the works, intercepted him with the eager inquiry:

"Is anyone killed?"

"Yes, yes!" said he, "We are all killed! Every one of us is killed!"

And it was some time before he could be convinced that he was not among the dead.

INCIDENTS IN THE DEVELOPMENT OF MOTORITE

Motorite consists of a compound of about seventy per cent. nitroglycerin and thirty per cent. gelatinated guncotton, the mixture being compounded in such a way as to form a tough and rubbery substance. This material is made into bars, which are smoothed and varnished upon the outside and then forced into steel tubes. In use, these steel tubes are placed in an apparatus in such wise that the bar of motorite can be ignited only at the exposed end, in a combustion chamber, into which water is forced, and as the combustion is confined to that end, it proceeds with absolute uniformity, according to the pressure, and without explosion. In other words, the motorite acts as a fuel, the products of combustion serving as a flame blast, blowing the water through a series of baffle plates, atomizing it, and converting it instantly into steam. The object of motorite is to replace compressed air in the driving of motors for self-propelled torpedoes.

I have already expended more than fifty thousand dollars in experiments with motorite and on different kinds of apparatus for its use. As about four times as much energy is available for driving a torpedo by this system as by any other, I hope some time to effect arrangements for the equipment of torpedoes with it.

The first bars of motorite that I made, I formed by passing through a die. The result was that a small, microscopic flaw which could not be seen with the naked eye extended through the bars from end to end, so that, when the bar was placed in the combustion apparatus, the flame of ignition passed immediately down through the flaw, exploding the apparatus.

After the first apparatus blew up, I made another one, and, as I could not very well conduct the experiments at the place where the first mishap occurred, I hired a floor in a building to make the test. But I needed an assistant, and it was problematical where I could find one.

One day, while returning home, I was accosted by a panhandler, a young man claiming to have just arrived from Pittsburg, seeking work. I told him that if he was actually looking for work I had a job for him, and I bade him come right along with me. I took him home that night, fed him, and watched him.

The next morning we went down to the shop. I explained to him all about the nature of the experiment that I was about to try, and told him that, if he had any timidity, the time to abscond was then and there. He told me that he was not afraid of anything, if I was not.

"Very well, then," I said, "I do not expect the thing to blow up; otherwise, I would not be here."

I got my time-watch ready, and told him to press the electric button to ignite the motorite. Instantly there was a terrific explosion. The windows were blown out into the street, and pieces of the shattered apparatus were driven into the ceiling and into the wall all about us; but fortunately neither of us was hit.

John looked calmly about him for a moment, and then at me, and remarked:

"God, she busted!"

While we were recovering from our amazement, half a dozen policemen rushed into the place, accompanied by a priest. I explained the mishap the best way I could, and, seeing that the priest was a handsome, genial, good-natured fellow, I appealed to him. He had a little chat with the policemen, and they all left.

I sent that priest a box of the best cigars that I could buy.

Under the counsel and advice of the landlord I then moved away from there.

I next bought a house, and in the back yard I built a laboratory with no windows or doors in it, except a skylight at the top and the windows and doors that fronted my house. The walls were of brick, and made thick. The skylight at the top was a large one, and was arranged to open up full size.

Special precautions were taken by means of various attachments to cause the roof to stay on in case of emergency.

A new apparatus was made and erected and got ready to test. This time my wife was my assistant, and we arranged to touch the thing off by electricity from the house. Again it exploded, and one of the fragments of the apparatus, coming through the open door, struck the wooden wall behind which my wife and I were standing and nearly passed through it.

On entering the laboratory, I discovered for the first time what was the actual cause of the trouble, namely, the longitudinal flaw already referred to, evidenced by the fragments of the motorite that remained after the explosion, for motorite, like smokeless gunpowder, when subjected to explosive pressure, is immediately extinguished upon the release of the pressure, so that when the apparatus blew up, all of the unconsumed motorite was extinguished, just as when the projectile leaves the gun any unburned smokeless powder is extinguished and is blown out in front of the gun, where the partially consumed grains may be recovered.

The next motorite was made by rolling the material into sheets, cutting into discs, sealing them together under pressure, and in that way building up the bars, which precluded the possibility of there being any flaws.

Some motorite was soon made in this manner, and another apparatus constructed, which was tested and which worked very satisfactorily.

Following this successful result, I built a laboratory at a dynamite plant near Lake Hopatcong for conducting the experiments on a larger scale.

My assistant at the motorite laboratory was one of that American country type, absolutely honest, perfectly fearless, painstaking and diligent, of such timber as the great men of the earth are made. He was altogether a most lovable fellow. He had all his life worked with explosives, and was a veteran in the manufacture and use of nitroglycerin and dynamite. But, when doing pioneer work with explosives, there is always an unavoidable element of risk, even when the greatest care is taken.

We at first had the hydraulic press, in which we built up the sticks of motorite, located in the laboratory room itself; but I suggested to my assistant one day that it had better be placed outside, and a heavy brick wall built between us and the press, as a barricade in case of a possible accident.

"For," I said, "suppose you should by oversight neglect to put in the leather packing between the piston and the motorite, we might have an explosion."

He said that he could hardly forget that precaution. Nevertheless, the press was placed outside, and the barricade built. The very first time that we used the press thereafter he did forget the packing, with the result that the press exploded. Although we were behind the barricade, still the concussion brought us to our knees. Had the explosion occurred while the press was being operated in the main laboratory, we should both have certainly been very seriously injured, if not killed.

It was a matter of several months before the full-sized torpedo apparatus with which we were to experiment was completed and erected, and the necessary quantity of motorite made.

On the day before the regular test was to be conducted, I was called to Morristown, as expert on a case in court, and I left orders with my assistant to make up an additional small quantity of sealing compound, used for sealing the discs of motorite together in building up the bars. This sealing material was made of a mixture of nitroglycerin, guncotton, camphor and acetate of amyl.

As I did not receive the telegram to go to Morristown until after I left home that morning, my wife expected that I would be working at the laboratory that day, but knew that I might possibly have a call to Morristown.

On my way home that evening, I was informed by a neighbor that there had been an explosion in my laboratory, that my assistant had been killed, and that the place had been burned down. I hastened to the spot and found my wife there waiting for me. All that was left of my assistant lay in an adjacent building covered with a piece of sacking.

That was one of the saddest moments of my whole life. It is impossible to know what little slip or misjudgment may have produced the explosion. A little inadvertence in the handling of a bottle of nitroglycerin may have been the cause.

The manner in which my wife was informed of the accident was about on a par with that employed by the Irishman who took the remains of a fellow-workman, killed by an explosion, home to his wife in a wheelbarrow, and, knocking upon the door, asked:

"Does the widdy McGinnis live here?"

She replied: "Indade, and I'm not a widdy."

And he said: "And faith ye are, for I have his rimnants here in the wheelbarry with me."

A butcher was the messenger-bearer to Mrs. Maxim. He said:

"Mrs. Maxim, have you heard the news about the explosion?"

And he continued: "Mr. Maxim's laboratory blew up and burned down today. They have found some of his assistant, but they haven't found any of Mr. Maxim yet."

Mrs. Maxim immediately rushed to the scene of the accident, where she learned the welcome news that I was in Morristown that day.

It was a matter of another year of hard work before I was again ready to make a new trial of the torpedo apparatus. There were several amusing experiences in connection with that testing.

The apparatus held a charge of one hundred and ten pounds of motorite. Water was pumped continuously through a water jacket over the steel cylinders containing the burning motorite and into the combustion chamber during each run. The apparatus was provided with an exhaust valve so constructed as to control, to a nicety, the pressure in the combustion chamber.

Under three hundred pounds pressure to the square inch, which was what was mainly used, the motorite burned at the rate of a foot in length per minute, and as each foot in length weighed twenty-five pounds, it burned at the rate of twenty-five pounds per minute. Each pound of motorite evaporated a little more than two pounds of water, and the products of combustion, mingling with the steam produced, escaped from the exhaust valve through an inch-and-a-half nozzle.

The roar of the escaping gas and steam was so great that it was impossible to hear one shout at the top of his voice. The loudest shout was less than a whisper. The roar could be heard with great distinctness more than two miles away. A good idea can be had of the violence with which the steam and gases escaped, from the fact that a door, which accidentally swung shut during one of the runs in front of the nozzle, although seven feet distant, was blown from its hinges, broken in two, and the fragments hurled twenty feet away. The noise was so confounding, that it was some time before my assistants and myself could keep our senses about us and note and record the pressures on the various gauges during a run, although the apparatus was separated from us by a barricade so strong and heavy that there was no possibility of our being injured, even should there be an explosion.

One day, just as we were about to make a run, the superintendent of a nearby explosives works called upon us, and I asked him if he would like to see the run, and he said he would.

I then asked him to note particularly and to record the pressure on a certain gauge. The run lasted about five minutes and, on turning to him for his notes, he himself was surprised that he had been so confounded by the noise that he had not thought of looking at the gauge at all.

On the day when the final test took place, the firm of torpedo-builders that was interested with me in the apparatus sent several representatives, including their chief engineer, vice-president and other officers of the company, to witness the test. Everything being in readiness, and each member of the committee being convinced that there was no possible danger in remaining in proximity to the apparatus and back of the barricade, while it was being tested, I gave each of the committee explicit instructions to watch the various gauges and to note the pressures, while the chief engineer and myself were to watch the nozzle gauge, and to observe the character and force of the steam and gases escaping from the nozzle.

I told the several members of the committee it was indispensable that they should carefully watch the pressure gauges during the entire run. As it was a condition of the test that I should get up steam within ten seconds, the chief engineer stood ready with his stop-watch when the electric button was pressed to ignite the motorite.

As the action was instantaneous, that is to say, as steam was got up practically instantaneously and was escaping at the nozzle under full head as quickly as a gun could be fired, he did not think of his stop-watch, and it was some little time before I could get him to look at the pressure gauge on the nozzle, so as to observe the character of the escaping steam. His eyes had a blank, meaningless look, but it must be confessed that he had the grit to stand there. Not so, however, with the other members of the committee. Each of them was far more interested in his own individual run than he was in the run of the apparatus, for not one of them was in sight when the run was completed. They came straggling back sheepishly, but no urging sufficed to bring them near the apparatus during any of the succeeding runs.

THE MULE GUN

In the old days when the Indians were sometimes troublesome on the Western frontier, an officer in the regular army, who was rather an ingenious fellow, conceived the idea of making a mountain gun out of a mule and the barrel of a common field-piece, using the mule for the carriage. He therefore had the gun securely mounted on the back of the beast.

They had not proceeded far with this novel battery, when a small knot of hostile savages was espied quietly eating their midday meal within easy range. The mounted gun was forthwith loaded heavily with grape and canister, the mule taken by the head and pointed in the direction of the Indians. A short piece of fuze that had been placed in the touch-hole of the gun was ignited.

The mule, hearing the sizzing of the fuze, began to rear and snort and kick and whirl about, while the officer and his men scudded to cover, and flattened themselves out upon the ground. They had not long to wait when there was a terrific crash. The gun had exploded under the overcharge, with the utter demolition of the mule carriage.

The Indians, hearing the report, looked quickly about them, and seeing the fragments of an exploded mule rocketing through the air, were frightened nearly out of their wits, and fled precipitately.

HOW GUSSIE GOT LOADED

When I was a young man I taught several terms of school in Maine, where, in the small country districts, the teacher is expected to be a walking encyclopedia of information.

One day there came a loud knock upon the door of the schoolhouse. On going out to see what was the cause of the imperative summons, I found standing there the wife of one of the neighbors, white as a sheet with agitation and alarm. She excitedly told me that her little boy, Gussie, had just swallowed a bullet, and she asked me what she should do for him.

"Why," said I pleasantly, "Give him a good charge of gunpowder. But be careful not to point him toward anybody."

She went home and gave him a dose of gunpowder, without ever seeing the joke.

DYNAMITE'S FREAK

A contractor, who does business up in New York State, told me the following story:

A carload of nitroglycerin dynamite had been shipped to him, but was held up in a freight-house for a day or two before delivery to him. One night while it was there, the freight-house took fire. Hearing the fire-alarm and looking out, he was astounded to see that it was the freight-house burning. Believing that his carload of dynamite would be sure to explode, he started to run to the scene in all haste, to warn the firemen and others to keep far away from the inevitable explosion, when, suddenly, there was a great burst of flame, which shot high into the sky and flared out bright and wild in all directions, sending up an enormous column of smoke. But this fierce combustion lasted only a few minutes, and then subsided.

He knew that his dynamite had burned up, and, curiously enough, without exploding.

He met the fire chief after the conflagration, and they spoke of the fire. The chief remarked that there must have been some very combustible freight on one of the cars. He said that, when the fire first started, the firemen played a full stream of water on this car, but it did not do any good. The car burned so fast and so fierce that they had to rush away for their lives, or they would have been consumed by the intense heat, and he wondered what it could be that would burn so fiercely.

When told that it was a carload of dynamite, he felt like a man who discovers the next day that he had, during the night, walked along the sheer edge of a high precipice.

Although dynamite in such quantities as a carload when ignited is almost certain to explode instead of merely burning, still, sometimes, even that quantity will take fire and burn up completely without exploding; while, at other times, a single stick of dynamite when ignited will detonate.

EXPLOSIVE VAGARIES

One of the old importers of picric acid in this country told me the following story:

He sold about five tons of picric acid to a manufacturer of dynamite doing business at a certain place up the Hudson, for employment as an ingredient in a particular kind of high explosive.

Not being very familiar with picric acid and the character of the exploder necessary to detonate it, the purchaser had poor success with it, and he called upon the importer with the grievance that he had been sold such a poor lot of picric acid that it was actually non-explosive, and was therefore practically worthless, and he wanted the seller to take it back immediately.

The importer could not convince him that he was mistaken, although he insisted that it was only necessary to know how to explode it, and that, when properly detonated, it was one of the most powerful explosives in the world.

"No," said the purchaser, "that picric acid you have sent me is not an explosive."

He admitted he knew that picric acid was recognized as a very powerful explosive; but he was sure of one thing—that the picric acid that had been sent him was not an explosive.

"Why," said he, "it is no more explosive than sand, and I want you to take it back."

"All right," said the importer, "you may return me a sample of it, and I will submit it to the requisite tests, and if it proves an inferior lot I will take it back."

That day, during the purchaser's absence, some workmen were moving a barrel of the picric acid in order to let a plumber mend a small leak in a lead pipe, which supplied the place with water. Over and about this lead pipe had been spilled a considerable quantity of picric acid, which had formed picrate of lead with the lead pipe.

The friction from the barrel set off this picrate of lead, which in turn detonated the picric acid; and the whole five tons went off with such violence as amply to demonstrate its explosive qualities.

The following day the purchaser returned to the importer with the complaint that that picric acid sold him was the most sensitive, most violent and treacherous explosive material in the world.

The importer laughed, and reminded him of their previous conversation. But, as the dynamite factory had been demolished and several men killed, the purchaser did not respond very readily to the humor of the situation.

THE TURKEY THAT WENT TO BED

Possibly it may not be diverging too much from dynamite stories to tell of an experience of mine with a steam-cooker, which I invented away back in the eighties.

In this cooker I was able to roast and bake by superheated steam. Sometimes it worked very well. At other times the safety valve gave me a great deal of trouble, being altogether too uncertain in its action.

One day I was sitting alone in the kitchen, steam-roasting a turkey, when dispossessed Bridget, who was waiting in an adjoining room, opened the kitchen door, and took a sly peep at me. I was endeavoring to convince her that the thing was perfectly safe, when, of a sudden, that cooker blew up. The kitchen windows were blown out, the door ripped off its hinges, and the stove demolished.

Fortunately, none of the fragments found either Bridget or me. The oven portion of the cooker, containing the turkey, went upstairs somewhere, through the ceiling. Later developments showed that the turkey had gone to bed in the room over the kitchen.

That cooker was my first patent.

BILL BENNETT, DETECTIVE

We had a neighbor, when I was a young man down in Maine, by the name of Bill Bennett, a hard-working farmer, who was very proud of his pile of dry hard wood, which he had prepared for the winter's cold.

Late in the autumn, however, the wood began to disappear faster than he thought it ought. He was sure that someone was stealing it, and inasmuch as his nearest neighbors had no store of wood whatsoever, and, too, were notoriously shiftless, he concluded that they must be the pilferers.

A little bit of detective work that he practiced to ascertain the truth of his conclusions was certainly ingenious and worked well.

Bennett took a dozen sticks of wood, and bored a large hole in the end of each of them, which he filled with rifle powder, putting about a pound into each stick. He then plugged the holes skillfully to conceal the evidences of his work, sawing off a short bit of the plugged end of each stick, so that the plug would not show, and distributed these sticks upon the part of the pile that was shrinking. He was careful to select the wood for his own burning from another portion of the heap.

The following evening he was looking from his window toward the house of the neighbors, wondering how long it would be before his ingenuity bore fruit, when suddenly there was a flash, a crash and a roar, followed by screams of "Murder!" and "Fire!"

The mystery had been solved.

WINNING THE OX

This Bill Bennett was a good deal of a marksman, and one day while attending a county fair, where he had imbibed a considerable measure of bottled-up unsteadiness, he came reeling along to a group of men who were engaged in shooting at a target. The range was long, and the price paid by each contestant for a chance to display his skill, or lack of it, was a dollar a shot, but the prize was a fat ox, which was destined to go to the first who made a bull's-eye. As yet none had succeeded in making the lucky shot.

Bill staggered into position, and threw down a ten-dollar bill for ten shots. The attendants steadied him sufficiently to confine his wild target practice to that part of the sky and horizon where the target was located.

Bill had wasted nine shots without coming within speaking distance of the target, which to his drunken sight appeared to be double. Rolling like a ship in a storm, Bill brought the gun to his shoulder for the last round, declaring, "By gum, I'm agoing to hit one of them targets this time."

And he did. As they went sailing by, he let blaze at them, and behold, it was a bull's-eye! Bill had won the ox on a one-to-a-million chance.

A DUEL TO THE DEATH

In the old pioneer days of Maine, when it was still a province of Massachusetts, a young French officer had an altercation with the chief of the Oldtown Indians, and according to the custom of the times, challenged the Red Man to fight a duel with him.

The old Indian, according to the courtesies of the game, was allowed the choice of weapons, and he chose two kegs of gunpowder. Each was to sit upon a keg, with the bung out. Then two pistols were to be discharged in succession. On the firing of the first pistol, two iron pokers, heated to a white heat, were to be laid upon a table beside the duelists, which was to be immediately followed by the discharge of the second pistol. At this signal they were each to seize a poker and thrust it into the bung-hole of the keg on which his adversary was sitting, the old Indian calculating that he would be quicker than the Frenchman.

But the Frenchman had a little calculation of his own, and he figured out something that the Indian had doubtless not thought of. This was that the explosion of either keg would be certain to explode the other.

But he made a bluff of it, thinking that the old Indian too might be bluffing, and so everything was arranged. Each mounted his respective keg and the first pistol was fired. The savage was a graven image, but the Frenchman did not wait for the second signal, and unlike Lot's wife, he never looked back.

THE BEWITCHED FLINTLOCK

My father used to tell a good story about a one-time chief of the Oldtown Indians, and, as it had to do in a way with explosions—indeed, a series of them—I add it to my collection.

There was a farmer living in an adjacent town, who frequently received visits from the old chief. On such occasions, the Red Man always carried his shotgun with him. The weapon, according to the times, was a flintlock, single-barreled muzzle-loader.

One day in the autumn, the farmer was feeding his turkeys by stringing a long line of corn upon the ground, on either side of which the turkeys were standing, head to head, in two opposing ranks for the feeding. The Indian was present, and the farmer asked his guest what he would give for a shot at that double line of turkeys' heads. The Indian answered that he would give five dollars, if he could have every turkey that he killed or wounded. The farmer, who had previously drawn the shot from the Indian's gun, leaving only the powder charge, accepted the offer.

The Indian leveled his gun and fired; but not a turkey fell. The old Red Man looked puzzled. The farmer laughed at his marksmanship, but the old savage merely grunted, and went home.

The chief appeared again next day, and the farmer asked him how he would like to take another shot, having again drawn off a charge of shot from the Indian's gun. He would gladly give another five dollars for a try. This time the discharge of the gun brought down a goodly number of turkeys. The Indian had taken the precaution of loading his gun with a double charge of shot. On the next visit received from the Indian, the farmer unloaded the gun down to the powder charge, then put in a wad of punk, and another powder charge with another wad of punk, and so on, until he had loaded the weapon nearly to the muzzle. He then replaced the gun in its position in the corner, dropped a fire-coal into the muzzle, and invited the Indian to supper.

After the lapse of a few minutes, the Indian's gun went off, bang! Much surprised, the Indian looked around, and remarked that it was a strange occurrence, that he had never before known his gun to go off by itself. While he was still cogitating over the strange occurrence, bang! went the old gun again.

The Indian hurried through his supper, very greatly perturbed, but he had not quite finished when the old gun spoke yet once again. The chief rose from the table hurriedly, seized his ancient weapon, and started off for home with as nearly a display of agitation as is permissible to the dignity of the Red Man. Before he had gone far, however, the old gun uttered another bang! He then broke into a rapid run, and just as he arrived at his wigwam, the gun banged again. Now thoroughly frightened, he hurled it from him over a fence. Still, for more than two hours the Indian's weapon continued its mysterious barking.

When the farmer explained the trick to the old chief, he felt that he had been somewhat compensated for the loss of his turkeys.

WHEN HE SHIRKED

A prominent financier, who was a much better business man than he was inventor, read of Moissan's experiments in making artificial diamonds. The financier conceived the idea of converting anthracite coal directly into diamonds by subjecting it to enormous pressure of gunpowder exploded in a strong steel cylinder.

As he wished to market a large quantity of his manufactured diamonds before their artificial character should leak out, he determined to conduct his experiments very secretly; consequently, he put the man-of-all-work at his country place upon the job. This faithful and useful servant was to report the progress of the work regularly at the city office of his employer.

After trying several experiments with black gunpowder, the man reported that the scheme didn't work—that no diamonds were produced.

The financier then told the useful that he had evidently reached the limit of power of black gunpowder.

"Now try dynamite," said he.

There was a break in the chain of reports, and he wrote the useful, asking him why he did not report. Still no answer.

After waiting some days, the idea suddenly struck the financier that possibly the process had proved successful and that the useful planned to betray him. He accordingly sent a peremptory telegram to him to report at once on pain of discharge.

The next day a vision, swathed and bandaged and perambulating on crutches, entered his office.

"You infernal old scoundrel!" yelled the wreck, as he entered. "Blow a man up with dynamite, and then threaten to discharge him for not reporting!"

THE ELEVATION OF WOMANHOOD

I had a certain man in my employ down at Maxim by the name of Benjamin Billings, whom we called Ben Billingsgate. Ben held views very strongly prejudicial to dogs and matrimony. He was all that is implied by the term "all-round useful." Though an erratic fellow, he was bright and energetic and seemed to be able to do anything under the sun when he set about it. But he lacked initiative, except in the expression of his opinions about those two abominations—dogs and matrimony.

When he was young and ardent he had married Sukyanna, a maiden who was dominated by the delusion that she had been born with a mission, to which all other considerations were secondary and should be subordinated. She was also a woman with a pug dog. Benjamin's nerves had been frazzling out for some time, and his patience was sorely tried by the division of the lady's affections between him and the dog—with a decided leaning toward the dog.

One day he brought home to his wife a beautiful Christmas present, which consisted of a large colored photograph of himself, mounted in an exquisite gilt frame. The expense of the thing represented a week's hard labor, but he wanted to create an impression upon his wife. He believed in doing things by wholes and in striking hard to win. His wife was very pleased—with the frame.

On his return from work the following evening, he took a sidelong glance toward the mantel over which the picture had been hung. He did not recognize himself. There in the frame was a life-size photograph of the pug in place of his, which Sukyanna had removed.

He uttered never a word, but his whole mental mechanism was turning somersaults. The next day, at roll-call, that dog was reported among the missing.

Benjamin pretended to sympathize and to condole with his wife, but she was disconsolate. Some Gypsies had passed that way during the day, and it was suspected that they might have stolen the dog. The horse was accordingly hitched up and a drive of ten miles was taken. When the Gypsies were overhauled and rounded up, the pug was not discovered. Then an advertisement was inserted in all the town papers. Still no pug. The canine continued a persistent absentee.

As a matter of fact, Benjamin had devoted ingenuity enough to the destruction of that dog to form the basis of a Sherlock Holmes detective story. He had prepared a sort of canister-bomb, adapted to go off by a strong thump of any sort. The dog, the bomb and a stout rawhide string, with which to tether the bomb to the dog, were confidingly placed in the hands of a small boy in the neighborhood, known to have both a sense of humor and a taste for the mischievous. The boy was, however, fond of dogs, and it eventuated that he decided to keep the dog for himself. Hence the delay in the finale of this story.

But the urchin's sense of humor finally got the better of his affection. He found it impossible to choke off the appeal to his imagination of hitching that bomb to the dog's tail. Consequently he took the pug out and carefully tied the canister to its tail. Following the ingenious instructions of Benjamin, as soon as he had done so he dodged into the house and shut the door before the dog realized what had happened.

When the pug discovered itself a part of an infernal machine, old-home-week associations rose up in its memory, and it made a bee-line for home and human mother.

Benjamin had made a little miscalculation about the amount of thumping that would be required to actuate the exploding mechanism of his ingenious bomb, and it did not explode immediately, as expected. The dog and bomb, consequently, hurtled through space like a comet with a head on both ends of the tail.

On the dog's arrival, Sukyanna was going about her household duties, with a book in one hand written by Miriam Mushroom on The Transcendentalism of the Universal, and Its Relevancy to the Elevation of Womanhood; while, with the other hand, directed only by subconscious mental process born of habit, she was preparing supper for Benjamin. She prided herself on that power of concentration and absorption, so common to the artistic temperament, which can resist for a while the battering-ram assaults on consciousness of howling children, barking dogs, or a house on fire.

As a result, she did not hear or see puggy as, with whine and din and clatter, he rushed into the room where she stood. Not receiving the expected attention and consolation, puggy in his impatience circled around the human mother, entwining the shanks of her in the strong rawhide cord, until dog and bomb had effectually hobbled her skirts, when, tripping, she went down on both.

This mean trick on the part of Benjamin bruised her artistic soul and proved far too much; she instantly separated from Benjamin—in the direction of the empyrean.

She had at last achieved the realization of the Elevation of Womanhood.

DIDN'T KNOW IT WAS LOADED

At the works of the Maxim-Nordenfelt Company in England, when some of the early experiments with smokeless powders were being made in that company's laboratory, a strong hydraulic cylinder, which had been employed for compressing experimental explosive materials, was thrown out of commission by the ram, or plunger, sticking in the cylinder. The cylinder was taken to the shop, and the job of getting the plunger out of it was given to one of the workmen. He thereupon commenced in his own peculiar way by heating the cylinder over a forge, thinking to expand it sufficiently to allow the plunger to be removed.

He succeeded before long, with an effectuality that perfectly dumbfounded his slow sense of expedition. The contained explosive naturally ignited, and the plunger was blown out like a shot from a cannon. The cylinder itself was blown downward, demolishing the forge, passing through the plank floor, and burying itself in the ground, while the plunger whizzed upward through the roof, and disappeared in the direction of Scotland.

THE WRONG TAP

The worker among high explosive materials must never relax his ceaseless vigilance. Not only his own life, but also the lives of those working at his side, hang upon the thread of infinite care. This fact is emphatically illustrated by an experience of my own, while conducting some experiments with a continuous process for making nitroglycerin which I had invented.

Orders were waiting, and it would take a week of constant labor on my part to complete the apparatus. I therefore crowded the week into three days, working constantly day and night, without a moment's sleep or rest.

I had thought out every detail of the process with the utmost care. I had tested every step, unit by unit, so I was confident not only that the process would prove successful, but also that it would be safe to operate.

On the forenoon of the third day, everything being at last in readiness, I now prepared to turn on the acids and the glycerin. I was well aware of the grim possibilities of my being killed, for if I had made a miscalculation or any wrong determination, I knew that my life might be the forfeit. I gave little thought to the likelihood of my being incautious due to the tremendous strain to which I had so long subjected myself. As it happened, I was so worn out that at the very outset I turned on the glycerin first, instead of the acids. My hand was actually upon the acids tap before I realized my error.

In that vital moment, some secret sense or instinct called back my wandering wits in the nick of time, and, shuddering, I dropped my fingers from the tap. Had I turned it on after the glycerin began to flow, I must inevitably have been blown to pieces.

"WHENCE ALL BUT HIM HAD FLED"

I have a literary friend by the name of Marvin Dana, who, although he was for years editor of the Smart Set, once failed in a bit of à priori perspicuity. Some Italians were blasting out a bit of rock at Landing for the foundation of a new bridge, to carry the roadway over the railroad in that village. They had just finished charging a big, deep hole with dynamite, and had lighted the fuze, when Marvin started to cross the temporary bridge with his usual measured stride of ever-conscious dignity. The Italians, who had withdrawn to a safe distance, seeing him coming, and they being unable to speak English, gesticulated wildly, and pointed excitedly in the direction of the blast under the bridge.

The littérateur concluded that there must be something extraordinary going on down below there—something quite worth looking at, and, walking directly above the blast,8 leaned over the bridge and looked down. Just at that instant the mine exploded.

He was, happily, unhurt by any of the flying stones and débris, but the knock-down argument of the shock from the blast convinced him that such carelessness on the part of those Italians, with never a guard to wave a red flag warning pedestrians, was, indeed, truly shocking.

BREAKING HIS NERVE

Just back upon the hills that rise up from the southern shores of Lake Hopatcong, there is one of the most important dynamite works in the country. James Wentworth began his labors there first as an errand boy, at the age of twelve, soon after the works started. It was his brag that he had grown up with the works, but that he had never gone up with them, although he had seen many another go up, when, on occasion, by some freak of chance, a packing-house or a nitroglycerin apparatus would be blown to the four winds of heaven, spraying wreckage of men and timber over the whole celestial concave.

Jim had no lack of courage. He had worked in every department of the business; had made nitroglycerin and nitrogelatin, and had become one of the most skillful dynamite packers. As he did piece-work, he made money rapidly.

One day, at a church strawberry festival, he was drawn into the vortex of that swirl8ing passion, love, and married. The young wife importuned him to give up the dynamite business, as he had already laid up sufficient money to start him in another business. Yielding to her wishes, he gave notice that his resignation was to take effect at the end of two weeks.

On the third day of the period of his notice, on the advent of the noon hour, he was seized with an uncontrollable impulse to take his dinner-pail and himself out of the packing-house where he was working. He said afterward that he got to thinking, "Suppose this packing-house should blow up; what would become of Susie?" — to say nothing of his own dispersion.

He went to the top of an elevation to eat his dinner, in full view of the packing-house, continuing his pessimistic reflections.

The place began to look suspicious. For the first time in his life he felt fear. On a sudden, that packing-house became a white, dazzling ball of flame, and he was knocked down by the concussion.

He told the superintendent that the three days he had served on his notice must suffice—he had lost his nerve!

THE GRIZZLY CANNON BALL

In the early days, when there was more individual and less corporate mining in the gold country of the West, a long and lean Yankee, Jim Evans, who was once a neighbor of mine in Maine, contracted the gold-fever, and went West.

Luckily, he almost immediately struck a pay streak high up the face of a cliff, where there was a wide shelf of rock that afforded a very convenient roadway for his use, as well as considerable area for the transaction of his operations.

Someone before him had started operations on the same site, but had become discouraged and quit, leaving a big steel tank, open at one end, lying upon its side, the open end pointing, like a huge cannon, over the mining settlement a thousand feet below. Jim used this tank to warehouse certain edibles, together with a keg of black gunpowder.

One day, on Jim's return from grubbing in8 the ground, he was amazed to find the entrance to his warehouse blocked by a huge grizzly bear that had crawled in to get at the edibles, and that fitted the big tube like a wad in a gun.

Jim was addicted to humor, and as there was a three-quarter-inch hole in the tank near the closed end right over the keg of gunpowder, the head of which had been removed, it occurred to him that he might make it somewhat interesting for the bear by lighting a piece of fuze and dropping it into the gunpowder. This he proceeded to do, and the bear proceeded to leave that tube after the manner of a cannon ball.

Hearing the report, and seeing a large volume of smoke, the townspeople, looking skyward and Jim-ward, were astonished at seeing a ton of grizzly hurtling outward from Jim's place and descending upon them.

On Jim's return to the village that evening, he was surrounded by numerous interrogators regarding the bear. "Oh," he said, casually, "I found the bear in my shack, and just threw him out, that's all."

THE JOKE WAS NOT ON THE CHINAMEN

When the Alaskan gold excitement was at its height, a couple of adventurous spirits, prospectors from California, had expended several months of precious, good old summer-time and exhausted their resources in an endeavor to locate pay dirt by sinking a shaft into a narrow table of land which jutted out from a high mountain near its base.

After thawing and grubbing and blasting through fifty feet of earth, with no gold in sight, they came upon solid ice underlying the cover of earth through which they had penetrated.

They kept on, however, for several weeks more, in an endeavor to penetrate through the ice; but they found ice, and only ice, for another fifty feet.

Then it was that it occurred to them to salt that ice with fine gold dust and sell out to some tenderfoot sucker.

They very easily found the desired victims in two Chinamen, with evident ample means and sufficient lack of experience.

The two prospectors had about a ton of dynamite on hand. This they lowered into the shaft and concealed it in a side drift just deep enough and big enough to hold it, calculating that the first shot fired by the Chinamen would set off the dynamite and, by completely demolishing the shaft, conceal their fraud.

The first blast made by the Chinamen did explode the dynamite, which not only wrecked the shaft, but also lifted the whole jutting bit of tableland—ice, earth, everything—sending it—an avalanche—down the mountain slope several hundred feet, exposing a thick stratum of glacial detritus, under where the ice had been, so full of gold that it proved to be one of the richest finds ever made in Alaska. The one blast had made the Chinamen millionaires.

CHINESE FIREWORKS

During the gold-digging days of California, before there was a restriction imposed upon the immigration of Chinese, a big American sailing vessel, while in Chinese waters, had taken aboard a large cargo of fireworks and a few tons of gunpowder of a special brand, which was safely housed in the hold, while all the sleeping quarters except those occupied by the crew, and all available deck spaces, were filled with a cargo of coolies to man California mines.

The vessel was one of those staunch, fast, sail-driven craft brought to their highest perfection in the shipyards of Maine just before the advent of the steamship.

When the ship was about a day out on its homeward voyage, the captain learned, through his faithful Chinese cook, that a big part of the Chinamen that he had picked up were half-breed Malay and Chinese pirates who had taken passage for the sole purpose of capturing the ship for piratical purposes, and that they were armed to the teeth, so that resistance offered by his crew of only twelve would be utterly hopeless.

While the captain was deliberating upon what to do, word was brought by his cook that the pirate horde were beginning to act very ugly, and had already taken possession of the fore part of the vessel, preparatory to a final assault upon the crew.

The captain ordered two lifeboats immediately to be filled with water and provisions and lowered, while he went below decks and lighted a train to the cargo of gunpowder and fireworks. Then the captain and his crew, together with the Chinese cook, manned the lifeboats and pulled away, to the amazement of the Chinese pirates, who seemed immensely pleased that they had captured the ship without a struggle.

The captain and the crew, in his two boats, lay on their oars at a safe distance quietly watching events, while the ship, which had now been turned about, was sailing away landward. When at a distance of about half a mile, that ship turned volcano. The whole above-water portion went up into the air with a belch of fire and thunder-roar like another Krakatoa, whose eruption shook the whole earth in 1883.

In their upward flight, Chinamen raced with rockets, while the heaven was filled with burning fireworks—and then it rained Chinamen. In fact, it was a real cloudburst of Chinamen, fire-crackers and ship's wreckage.

BROWN, THE GUNNER

For many years, all inventors and manufacturers having occasion to attend experiments with their productions at the Naval Proving Grounds at Indian Head, were aided in their work by Brown, the gunner. He was a very ingenious, genial, gigantic fellow, one of the most likable men in the world. There was nothing about the mechanism of guns and gunnery unfamiliar to him.

Once, during the early years of his service there, a fragment from an exploding gun struck him in the forehead, leaving a great dent. As soon as he recovered, he returned to his duties undeterred, although he had had many other close calls.

One day, a few years ago, he walked in on me at my summer home on Lake Hopatcong. During his visit, he asked me if I believed in presentiments. He said he had had a very strange presentiment of impending danger in his work at Indian Head. He told me that he had confided this to the commanding officer there, who laughed at him, and said, "Oh, Brown, at last you are losing your nerve. Go and take a two weeks' vacation, and then come back."

Brown did go back at the end of his vacation.

A few weeks later, while testing a new heavy gun, something went wrong. The breech block blew out, and Brown was killed.

THE HAPPENING OF THE UNEXPECTED

Some time ago, a young lady who had been my private secretary for about four years got married. Thinking that one of the best ways of securing another competent stenographer and typist to take her place would be to go to an employment agency, Mrs. Maxim and I called upon the manager of one of those institutions.

Mrs. Maxim, according to the habitude of her sex, led in the conversation. She told the manager about the unusual requirements that the person engaged must have—that she must have a good general education, must be very expert as stenographer and typist, and above all, must be an exceptionally good speller. Furthermore, Mrs. Maxim placed especial emphasis upon one stipulation—that we did not want a girl under twenty or a woman over thirty-five, for the reason that a girl under twenty is very apt to lack the necessary experience and serious-mindedness for such a position, while a woman around and above forty is apt to be set in her ways, and to lack the necessary flexibility of mind and nature readily to adapt herself to anything to which she has not always been accustomed, and is, furthermore, likely to be unable to learn anything new with the facility of a younger person.

The manager was all suavity, pleasant manners and promises, and assured us that he had on his waiting list a number of young women who would exactly meet our requirements, and that he would send three of them over that very evening.

We learned from the bit of experience which followed that employment agencies and those who are sent by them to apply for positions, are apt to be governed by reasoning similar to that of the small boy, who, seeing an advertisement that twenty-five dollars' reward would be paid for a Pekinese spaniel, thought it would do no harm to try, and so he called to claim the reward with a huge mongrel—a cross between a Newfoundland and a St. Bernard.

Well, at the appointed hour, two archaic dilapidations wafted themselves in upon us, who looked as though their nascency had a priority on the

Stone Age and they had been vouchsafed to us among the antediluvian survivors of Noah's Ark.

The first one—a slip of a girl of some sixty-seven to the nth-power summers and as many winters—betrayed her lack of typistical experience by mistaking a national cash register for a typewriter. Then she confided in us the little confidence that she really knew nothing about typewriting as yet, but that, in the sweet long ago, in the days of auld lang syne, she used to drum quite a lot on the piano, and, consequently, she imagined that typewriting, being a sort of mere finger play, would come so easy to her that she would have little difficulty in acquiring the necessary aptitude on a typewriter to qualify for the position.

The next applicant was a tall, slight, sinuous, willowy, sylph-like and ethereal creature of the hippopotamus variety, who floated into our presence like a breath of old winter, made sweet summer by the mingled odor of violets, lilacs, musk and new-mown hay. I gave her a short dictation, which she took down in longhand. I asked her why she did not write shorthand. She said she did write shorthand, unless she was in a hurry. Contemplating her huge bulk, I insinuated that we should want someone a little lighter on her corns than she, as one of the desirable accomplishments in a private secretary was that she should be able to play tennis. She said that although she had never played tennis herself, still it ran in the family, because her grandchildren were expert tennis players.

When the third antique entered, the thing began to get monotonous, as Mark Twain remarked, when a mule had fallen through his tent three times in one evening. We were getting out of patience. I told the old lady at once that we did not want anyone under twenty or over thirty-five. She assured me that she was not under twenty. I told her that I had guessed as much, and asked, "How about the other limit!" She sharply retorted that she had never, in all her life, touched thirty-five. "Well," said I, "if that be so, you must have been skidding some when you went by that numeral."

Disappointed, and highly indignant, we called again the next day upon that manager of the employment agency. He was profoundly apologetic, and said that he happened to have waiting in another room a young lady who was exactly what we wanted. She was immediately asked into the private office, where Mrs. Maxim and I examined her. She was about twenty-five years of age, and was, as they say down in Maine, as smart as a steel trap. I gave her a dictation replete with multi-syllabic terminology, and with unusual words of difficult orthography, but she took down everything with lightning speed, read back her notes to perfection, and transcribed them rapidly on the typewriter without a mistake.

We asked for what salary she would be willing to come to us. The salary asked was pretty high, but we instantly agreed to pay it. The manager and the young lady exchanged glances, and both looked a bit surprised. Mrs. Maxim and I then asked if we might talk with the young lady alone for a few minutes.

After some Sherlock Holmesy talk with the young woman, Mrs. Maxim and I came to the conclusion that she was a show girl kept by the manager merely to prove that he had the goods when required, provided anyone wished to pay a sufficiently high salary, and the salary was made high enough to deter most applicants. We got it from the girl that she had several times been hired and had worked a few days for each of a number of employers, until she could find some rational excuse for breaking away and returning to the agency, the manager of which, we also learned, was her brother, and she was a partner in the business.

The incident reminded me of a story told by a friend of mine in New York who bought a beautiful and highly trained Scotch terrier of a Broadway dog vendor, thinking that after keeping the dog tied up for a week, feeding him and treating him with kindness, he could be depended upon to stay with his new master, but the moment the dog was freed he disappeared, and the next day he was again with his master, the dog vendor, ready to be resold. Some time later, a light was thrown upon the inner consciousness of my friend by reading an account in the newspapers of the arrest of the dog vendor for obtaining money under false pretenses and practicing fraud in the sale of dogs, or rather, of the dog. The canine was a sort of homing-pigeon dog, trained, like a carrier pigeon, to return from each new master as soon as freed. The buying and selling of that one dog constituted the main business of the scamp.

When our interview with the young woman was concluded, we started to leave the office in disgust, but at that moment a young woman of rather prepossessing appearance, about thirty years of age, entered the office looking for a position. She explained that her late employer having gone to Europe, she was looking for a new place.

After a critical examination, we found that she would meet our requirements very well. Then it developed that, having read in newspapers and magazines some of the accounts, highly colored by the writers of them, of how I cooked with high explosives and lighted my cigar with a stick of dynamite, and burned nitroglycerin in a lamp to light the room, she, being of a rather nervous temperament, was afraid of the prospective companionship with explosive materials.

I assured her that the accounts were misrepresentations of actual facts, and explained that we lived at a very safe distance from any explosive works, and that she would be exposed to no danger whatsoever. I finally convinced her that our home was a safe place, and although still harassed with some doubts she decided to come with us.

In the edge of the evening, after her arrival, she and I were sitting at the dining-room table engaged in conversation. I was telling her how groundless had been her fears, when there came a terrific explosion. The sky was lighted up with a brilliancy that would shame the noon-day sun, and fragments of brands from the burning fell all about the house.

I confess that I was as much surprised as she was—and that was going some. I rushed out, and found that my tool-house, located about a hundred yards from my residence, had blown up, and the wreckage was on fire. Being sure that there were no explosives in the building, I was greatly puzzled.

There were in the place at the time perhaps a hundred rounds of Mauser rifle cartridges. These were exploding, one after another, from the heat. The neighbors who had run to witness the fire, were greatly frightened, and did not dare to render any assistance in putting out the flames, especially while the cartridges were exploding.

I ran to a hydrant nearby, got out the fire-hose, and found, to my amazement, what one usually finds under such circumstances, that the nozzle of the hose had been taken off, and the hose disconnected from the hydrant, and that there was no wrench there. I ran and got another hose and a wrench, made the connections, and ran out the hose to extinguish the fire, when I found that only a small stream of water as big as my thumb flowed from the hose. I then ran down to my house to see if there were any faucets open which would reduce the pressure, and then to the pump-house to measure the water in the supply tank, and found that the tank was nearly full, and that thirty-five thousand gallons of water were available for extinguishing the fire. Yet I could get no pressure. The result was that nothing was saved, and the building and all its contents were a complete loss. As there was no insurance, the loss was about fifteen hundred dollars.

After it was too late to save the building, I walked down to the Hotel Durban, on my property, which I supplied with water, to calm the fears of some of the guests who were agitated, when, to my amazement, I found a two-inch fire-hose turned on full, and running in the road. I learned then that a stupid fellow who was staying at the hotel, had turned the water on at several fire hydrants to play water on the hotel, although the hotel was at such a safe distance from the tool-house that there was not a particle of danger whatsoever. It never occurred to him to close off one hydrant when he opened another; consequently, the pressure was reduced so that no water at all could be had at the scene of the fire, and not pressure enough on the hose-pipes that he had turned on to do any good even had they been needed.

After things had quieted down, I returned to the house to resume my conversation, and to repeat my assurances to the young lady secretary, but I found a polite note tacked to the table-cloth, requesting that her trunk be forwarded the next day. She had not waited for further conviction as to the safety of her new position.

On investigation, I learned that a fire had started in the tool-house from some cause unknown, and had proceeded long enough to get one side of the interior of the building well ablaze. As there were five gallons of denatured alcohol in the place, and the same quantity of gasoline, and about ten pounds of sulphuric ether, it is probable that one of these had become heated and, bursting, set free a lot of vapor which, mixing with the atmosphere, exploded. There were also in the building about thirty pounds of finely pulverized aluminum, ten pounds of magnesium powder and other ingredients for flashlight powders, with which I intended to conduct experiments. As these materials were not mixed, they were not explosive, but their combustion was what produced the wonderful light when the explosion occurred. The result was not like that from an explosion of dynamite, in which case the building would have been literally blown to fragments, but, as is usual in gas explosions, the roof of the building was lifted up, the sides thrown out, and the roof dropped in. Even the front door of the building, charred from the initial fire, was found otherwise intact.

While sitting on the porch of my house on Lake Hopatcong, dictating this story to my stenographer, and when I had arrived at this point, she suddenly called to me, "Look!" pointing her finger across the Lake to a huge column of smoke going up from the Atlas Powder Works, and mushrooming out into the sky. The direct distance is about three miles, but it seemed quite a long time before we felt the shock and heard the sound. Although the sound was loud and the shock considerable, the sound was much louder and the shock much heavier even at longer distances in several directions, owing, I imagine, to the difference in the underlying strata of earth.

As I learned later, the explosion took place at one of the packing houses, which carried another packing house with it, together with a nitroglycerin storehouse, so that about ten tons of dynamite, or its equivalent, went up in that column of smoke. I understand that seven men were killed, and about twice as many injured. It was the largest and most destructive explosion that had ever occurred at those works.

WHEN THE WASH VANISHED

I was once invited to speak at a County Fair at Pittsfield, Massachusetts, where I used to live when in the publishing business. My subject was Explosive Materials and Their Use in Warfare.

The management was especially desirous that I should give my auditors some sort of spectacular demonstration, to show what explosives would do. A platform was erected in an open field, and I had an arena roped off at the rear of the platform about fifty feet wide, and running back several hundred feet. In the rear portion of this arena I buried several sticks of dynamite, and connected them with an exploder and a battery on the platform.

Also, I brought several cotton bosom-shirts, several cotton undershirts, half-a-dozen handkerchiefs, a couple of towels, half-a-dozen pairs of cotton socks, and as many cheap cotton collars and cuffs. These I had immersed in a concentrated mixture of nitric and sulphuric acids, converting them all into guncotton. Then I washed and soaked the acid out of them, and dried them.

I stretched a clothes-line from the speaker's platform to a distance of about thirty feet to my right, and on this I hung my guncotton clothes, only a few feet away from the front of the audience.

There were, perhaps, a thousand people massed in front of me, crowding up close, that nothing should miss them. I made a brief talk on the nature and use of explosives, and burned some smokeless powder under water, and then I touched off the dynamite in the rear of the field, which made a very pretty showing.

The audience was very curious about that wash. That I should have hung my linen out to dry on that occasion they thought was very peculiar taste, to say the least; and some of them did not hesitate to say that they considered it very bad taste.

I then said to the audience that I must beg their pardon for displaying my underwear as I had done; that I appreciated the fact that it was an unsightly display, and, to accommodate them, I would immediately proceed to get it out of sight. I then touched it off with an electrical igniter, and that laundry disappeared in one great bright flash of flame.

There happened to be in my audience an ingenious fellow with some knowledge of chemistry, who was a noted wag and practical joker. Taking the hint from my nitrated laundry, he nitrated a cotton handkerchief and sent it to the Chinese laundry with the rest of his wash.

When he called for his clothes, he found John Chinaman with his right arm in a sling. However, John was all smiles, and apologized for the absence of the one handkerchief, but said nothing more about it.

A short time after the fellow had put on his clean underwear, he developed a very severe case of prickly heat, followed, a little later, by a sensation like that of needles being stuck into his body over the entire surface. Anyone who has taken a bite of a wild Indian turnip knows what that sensation is. The Chinaman had charged his customer's garments with a preparation extracted from a Chinese variant of the Indian turnip. It took a couple of weeks, with the aid of a physician, for the wag to recover from the little unpleasantness which the Chinaman had inflicted upon him.

THE FRIGHTENED FISHERMAN

When testing big guns at Sandy Hook, the officers are often greatly annoyed by fishing boats that persist in getting within range of the guns and in remaining there, entirely regardless of the work or wishes of the officers of Uncle Sam.

It is a curious circumstance that, according to the law of the country, these ships have the right of way, and even the officers of the Government Proving Grounds have no power to compel a fishing smack to move out of range.

There was one boat of this kind that persisted in anchoring daily exactly in the range of a ten-inch gun that was under test, and day after day the tests had to be delayed.

One morning, however, there being a haze or fog floating close down upon the water at a distance of a couple of miles from the shore, and the sea looking perfectly clear to that distance, the officers in charge of the testing of the gun concluded that the range was clear, and they fired, but the captain of the fishing boat above referred to happened to be on his job, just as usual, though concealed by the fog. He had stretched himself out in a hammock on deck and was taking a snooze, when a ten-inch projectile passed through his boat under him, and ricocheted on out to sea. He kept out of the gunner's range after that.

Following this incident, one of the officers conceived the brilliant idea of keeping fishermen from coming into the line of fire in the following manner: When a boat was seen sailing into range he would fire several six-pound shells, exploding them in the water along the line of range, and directly in the path of the oncoming boat. This method served the purpose admirably. While the fishermen would calmly cast anchor and occupy a position directly in range of a gun being tested, they did not dare to sail directly into the line of fire of exploding six-pound shells.

THE COLONEL WAS PROVOKED

An Army officer tells me the following story:

One time, while he was on duty at the Sandy Hook Proving Grounds, they were testing a gun-shield to see whether or not it would resist the penetration of a six-inch shell.

The officer whose duty it was to attend to the loading and firing of the gun did not always allow the required time to elapse after sounding the warning before discharging the gun, especially when he took it for granted that no one was in the zone of danger, in which case he was apt to consider the signal of warning a mere formality.

Such was his attitude and action on the occasion to which this story refers: He gave the signal, and immediately fired. The projectile, which was expected to penetrate the shield, went only half through, and stuck there, when, to the horror of all participants, especially of the careless officer referred to, the Colonel of Artillery emerged from behind the shield, unhurt, but madder than a demon in Dante's Inferno.

No more guns were fired without the lapse of an ample period of warning.

WHEN THE DARKIES TURNED PALE

At one of our Government proving grounds, some years ago, the officers were testing a new high explosive, and, as was their custom, they charged a twelve-inch shell with the material in order to estimate the power of the explosive by the fragmentation of the projectile when the charge was detonated.

They had a bomb-proof chamber prepared for this purpose. It consisted of a room about ten feet wide, twelve feet long, and eight feet high, lined with armorplate. The projectile was placed on the armored floor in the middle of the room, and covered with a few hundred pounds of fine sand. It was armed with an electrical exploder, which was set off from another bomb-proof at a safe distance. After each explosion, the fragments were sifted from the sand and counted and weighed.

A twelve-inch shell charged with Maximite and exploded at Sandy Hook during the tests there of that explosive, was broken into ten thousand fragments. The fragments made deep dents in the hard face of the armorplate. The shell that enters into this story was exploded under similar conditions.

When the officers were ready to explode the shell, they sounded the usual alarm to give warning to laboring men on the premises to seek cover. Now, it so happened that about a dozen negroes who were engaged in some pick-and-shovel work had been in the habit of using this very bomb-proof as a shelter when a big gun was fired; consequently, when the warning was sounded, they immediately rushed for cover within that bomb-proof.

The officer in command was about to close the switch to explode the projectile, and his hand was already upon it, but, being an exceedingly cautious man, he thought he would take another look to be sure that all was safe, and, to his amazement, he saw a negro who had been screening himself behind a pile of rubbish making a dash for the bomb-proof containing the projectile, when it was revealed that the dozen darkies had all huddled into it for safety.

When those darkies found out how close a call they had had, they turned just as pale as negroes can turn. Had the projectile been exploded while they were in the bomb-proof, they would not only have all been killed by the blast, but would also have literally been blown to ribbons.

THE DOG THAT WAS A REAL MASCOT

In the long line of trenches that constitutes the French and British front, facing the equally long German front, the soldiers relieve time's tedium by numerous artifices. Many kinds of pets—dogs, cats, owls, doves, parrots—are harbored for the sake of their company, or as mascots—bringers of good luck.

A French soldier had a dog that was a great favorite in the trenches, for the reason that he was a famous ratter, and as the trenches were infested with rats, he was a most welcome guest.

One day, when the Germans were bombarding the French position before Verdun preparatory to a charge, a huge howitzer shell, penetrating deep into the earth in front of one of the French trenches, and exploding, buried half a hundred men—among them the owner of the dog.

The dog also was quite buried by the explosion, but he quickly dug himself out, and then he began an eager search for his master. Smelling out his location, he dug furiously with all his might to unearth him. Fortunately, his master's head was near the surface of the ground, but his arms and legs were bound tight so that he could not move, and he was nearly suffocated when the dog succeeded in digging out his head and face so that he could breathe.

Happily, relief came soon, and when the rescuing party arrived, they found the dog still working with all his strength to uncover his master.

Pick and spade soon brought the dog's quarry to the surface, who was quite unharmed except for a few bruises, while the dog, it was seen, was bleeding at ears, eyes and mouth from the effect of the explosive blast.

WEARY WILLIE'S DISCOMFITURE

Some good old English folk whose prosperous son had made a large amount of money in the railroad business in America, were persuaded by their boy to give up their fine, old-fashioned English country home for such home life as America could afford.

The dutiful son had anticipated the wants and pleasures of his parents, and on a fine country estate he had built practically a replica of the old English homestead. There was the big fireplace and the big, wide chimney, to be swept by the smutty chimney-sweep. The chimney was provided with pegs to climb up and down.

Some time after the good parents were quartered in their new home, Weary William the wanderer, a real hobo, walking past the place late one night, could see enough of it in the moonlight to recognize its genuine English aspect; for Weary Willie had, in his boyhood days, been one of those smut-faced chimney-sweeps in old England, and when he walked up and peeped through the window and saw a few embers in the familiar fireplace, he concluded to go down that chimney and take a nap in the cosy comfort that the room provided, and perchance find something to eat and drink without waking anyone.

Entering the room by way of the chimney, he did find, all set as though for himself, edibles and wine—left-overs from someone's late supper.

After feasting, he took a snooze on the sofa, intending to take his leave the way he came at an early hour before the family was up, but he had drunken more of the good wine than he ought, and he slept soundly. He was awakened by voices, which told him that it was high time and past for him to make his exit, and he scooted up the chimney in great haste, but not a whit too quickly, for by the time he had raised himself up out of sight, several persons entered the room. He did not dare continue his ascent or move for fear of making a noise. He waited there, breathless, for a more favorable opportunity to climb out.

It so happened that an ingenious Yankee neighbor of the English gentlefolk had suggested a more expeditious way of cleaning the chimney than by sweeping it out in the old British fashion. He said that all that was necessary was to throw several pounds of black gunpowder into the fire, which, flashing, would blow the soot out of the chimney. Of course, the genius had never tried the experiment himself, but as such geniuses are usually cocksure, he was so confident of success that he did not feel the need to make any preliminary experiments. Therefore, just as the tramp had mounted above the line of vision into the chimney, the genius, entering the room, threw the gunpowder into the fire, which instantly exploded with a great flash and smoke, blowing cinders and embers all over the room and filling it with dense, black, sulphurous smoke, burning the face and hands of the genius considerably, and frightening the elderly people out of their wits. But what frightened them all still more, was the appearance of the thoroughly singed and scared tramp, who fell from his perch in the chimney, down into the fireplace, and rolled out into the room, sneezing, coughing and saying things, all at once.

The terrified tramp was easily secured, and when the master's gold watch—a gift from royalty and a family heirloom—was found upon his person, the genius was not only forgiven for his miscalculated experiment, but also thanked for his good offices.

LO, THE POOR INDIAN!

Dave King, editor of the Morris County Press, Morristown, New Jersey, was reared a lariat man in the Wild and Woolly, in the days before civilization, rum and guns had subdued the Cheyennes, the Comanches and the Sioux to extinction or to the more uncongenial fate of enforced good behavior.

In all of Dave's hair-ruffling experiences—corralling stampeding long-horns, lassoing and riding a bull-buffalo bare-back, hunting, with Rex Beach, the great Kadiak bear in Alaska, whose enormous bulk and Ivan-the-Terrible disposition would by comparison make the grizzly of the Rocky Mountains a gentle companion—his most intimately interesting, close-to-nature adventure was when he was ten years old, and dwelt upon the upper waters of the Arkansas.

Dave's father, a husky pioneer, accompanied by his ten-year-old son, his brother, "Uncle Joe," an assortment of dogs, guns and ammunition, embracing a dozen kegs of gunpowder, had gone there to stake a squatter's claim, hunt buffalo and grow up with the country.

Timber was scarce, so, after the manner of the troglodyte, they burrowed out a room in the side of a hill, which constituted at once cook-room, dining-room and parlor, and also museum of rare weapons, dog-kennel and powder-magazine. The cook-stove was placed in the middle of the room, and the flue was run up through the ground for ventilation and the escape of products of combustion.

One day, Dave's father and Uncle Joe went on a buffalo hunt, much to the disconsolation of Dave, who wanted to go along. Toward the end of the afternoon following the departure of the hunters, Dave built a roaring fire in the stove to keep himself company, and incidentally to prepare supper for himself and the hunters, who were expected to return before sundown.

His eyes regarded longingly a double-barreled shotgun hung on the wall. He had many times been warned by his father to exercise caution in handling the guns during his absence, but Dave had the dare-devil spirit of his parent, with the added impulses of the small boy, and he took down the shotgun and fondled it lovingly, examining its firing mechanism. Then he proceeded to return it to its hanging, not noticing that he had left one of the hammers cocked. He did not know that the gun was loaded, and he would not have been deterred had he known. In putting up the weapon he accidentally touched the trigger of the cocked hammer and the charge in that barrel exploded, sending shot and burning wads under the sleeping-bunks, just missing one of the kegs of gunpowder.

Dave proceeded with his cooking, but soon he smelled smoke, and looking under the bunks discovered, to his horror, that a fire had started. Under the bunks he went, pawed at the fire with his hands, and smothered it with his hat, until he thought that he had extinguished the last spark. Then he started for a water-hole an eighth of a mile distant, to get a pail of water, accompanied by his favorite dog.

When he got out into the open, he saw a dozen horsemen just coming into view over a rising ground between him and the sinking sun. He thought at first that his father was bringing company home to dinner, and he waited and watched. But he soon saw by the feathered and blanketed make-up and demeanor of the horsemen that they were savages on the warpath.

Dave was not long getting himself and his dog out of sight in a badger-hole which he had, during many days of hard labor, enlarged for a playhouse.

The Indians were a party of Cheyennes who had been forcibly located in the Indian Territory by the Government. On this occasion, half a thousand of those fierce warriors decided to go on the warpath and return to their former hunting grounds in Wyoming. On their way they burned houses and slew and scalped everybody that fell in their path. Among many other outrages they, for a little diversion, killed and scalped a young woman school-teacher and forty pupils. United States troops then rounded them up and corralled them in Fort Robinson, Nebraska. One night they made a break to escape and the soldiers, now out of patience, killed the whole bunch.

But to return to Dave: When the Indians saw the smoke coming out of the top of the ground, their curiosity was excited, and discovering that it was a dwelling they rode round it, red-man fashion, in a constantly narrowing circle, firing guns and war-whooping.

The dog began to bark and struggle to free himself to get after those Indians, but Dave thrust his hand into the animal's mouth, and grasping his lower jaw managed to keep him from barking. It took all of Dave's strength to hold that dog, but he knew that it meant life or death, for if the dog should escape he would betray their hiding-place.

The Indians, finding no sign of life in the dugout except the barking dogs that Dave had shut in, came closer and closer. Half a dozen of them got up on the top of the dugout, and the others bunched themselves in close to the entrance, preparatory to rushing the place.

But Dave had not succeeded in extinguishing the last spark of the fire that he had started under the bunks, so, coincidentally with the Indians arranging themselves about the cavern, the twelve kegs of gunpowder went into action.

Dave could not imagine what had happened. He thought that possibly the Indians had captured the gunpowder and exploded it purposely, but he did not dare to emerge from his hiding.

There was an interval of silence. There were no more war-whoops, and he concluded that the Indians had departed. They had, but not exactly in the manner that Dave imagined.

The parent and Uncle Joe, returning on the edge of evening, were dumbfounded at finding only a great hole in the ground where the dwelling had been. Dave's father wrung his hands and bemoaned the loss of his boy, while Uncle Joe consoled him with the usual I-told-you-so that he ought not to have kept the gunpowder in the place.

They began a diligent search for any souvenirs of Dave that might have happened to return to Mother Earth. After they had gathered up about a wagon-load of the disintegrated Indians, Uncle Joe suggested that they must be on the wrong scent.

At this puzzling juncture, Dave, hearing the voices of his father and Uncle Joe, cautiously emerged from his hiding. When he came in sight, Uncle Joe said, "There's Dave now! There's your boy!" His father looked blankly at him for a moment. Though the vision looked like Dave he could not trust it. He said, "No, it can't be my boy! It can't be my boy!"

But it was; and Dave is still with us.